Results and Problems in
Cell Differentiation

A Series of Topical Volumes in Developmental Biology

4

Editors

W. Beermann, Tübingen · J. Reinert, Berlin · H. Ursprung, Zürich

Developmental Studies on Giant Chromosomes

Edited by W. Beermann, Tübingen

With contributions of

M. Ashburner, Cambridge · W. Beermann, Tübingen
H. D. Berendes, Nijmegen · M. Lezzi, Zürich
R. Panitz, Gatersleben · C. Pelling, Tübingen
D. Ribbert, Münster · M. Robert, Saarbrücken
G. T. Rudkin, Philadelphia

With 110 Figures

Springer-Verlag New York · Heidelberg · Berlin 1972

ISBN 0-387-05748-X Springer-Verlag New York Heidelberg Berlin
ISBN 3-540-05748-X Springer-Verlag Berlin Heidelberg New York

© by Springer-Verlag Berlin · Heidelberg 1972. Library of Congress Catalog Card Number 74-189387.
Printed in Germany. Typesetting, printing and bookbinding: Brühlsche Universitätsdruckerei Gießen.

Jack Schultz

May 1, 1904 — April 29, 1971

So great was the impact of the "rediscovery" of polytene chromosomes on the genetic study of Drosophila that MORGAN, BRIDGES, and SCHULTZ began their 1935—1936 report to the Carnegie Institute of Washington with the words:

"During the past two years the emphasis in the Drosophila work has shifted from analysis by breeding methods to analysis by cytological observations".

T. H. MORGAN had returned to embryological studies, interrupted 25 years earlier by Drosophila, and the report was a summary of the work of the other two authors. CALVIN BRIDGES was preparing detailed chromosome maps and cataloging the breakpoints of genetically known chromosome rearrangements in order to correlate his genetic maps with the new cytological ones. JACK SCHULTZ, among other projects, continued his studies on position-effect variegation. He noted variegation of the morphology of the salivary gland chromosome regions known to contain the genes for which mottled effects had been observed.

Polytene chromosomes continued to play a prominent role in JACK's thoughts and work for the rest of his life. Thirty years later in the 1965 Brookhaven Symposium we find original work supporting the earlier view that variegation resulted from the inactivation of genes in some cells but not others during development: in this case the development of a puff served as the indicator of activity. The intervening years had included a variety of studies aimed at understanding the phenomenon in terms of the regulation of the synthesis of nucleic acids, beginning in an era when proteins were known to have the range of specificities required of the genetic material.

But personal involvement in experimental work directly concerned with polytene chromosomes does not reflect the extent of his influence on matters discussed in this volume, namely the nature and function of the genetic material. His most important writings were preparations of critical analysis of data at hand (of which the 1965 symposium paper was the last and one of the best), indicating where evidence was strong and where weak, but, most important, drawing on a wealth of disciplines varied enough to avoid pitfalls so often associated with "specialization". It was the analysis that appealed to him and that he offered; those who looked to his writings for answers were sometimes disappointed by a carefully worded statement balancing evidence against interpretation, extracting the significant from the trivial, reluctant to select a clear single picture when conflicting evidence clouded the view. It was his ability to see the essentials and to project novel solutions based on a an unusually broad store of knowledge, subject to immediate recall, that made the flow of ideas communicated by personal contact more influential than the thoughts that were, from time to time, frozen in print. To this, most of the authors of this volume can attest.

GEORGE T. RUDKIN

Preface

Earlier titles in this series have directed attention to selected questions of current theoretical or practical interest in developmental biology, without special emphasis upon individual biological systems. Tho choose giant polytene chromosomes as the common denominator of a new volume may at first sight seem to set rather arbitrary restrictions upon a coherent discussion of the basic problem of cellular differentiation. However, the reader will find out for himself that, on the contrary, such a limitation, though dictated by the material, focuses the analysis upon one essential point: the differential release of genetic information from operational subunits of the eukaryotic chromosome.

In the organization of this volume we have not striven for completeness — some, indeed, may feel that the work on Sciarids is underrepresented — but the various contributors have made every effort to render a thorough and up-to-date picture of the situation in their particular field, and on this basis to discuss all possible angles of general biological interest.

Authors and editors have agreed to dedicate this volume to the memory of JACK SCHULTZ.

Tübingen, Berlin, Zürich   W. BEERMANN, J. REINERT,
June 1972   H. URSPRUNG

Contents

Replication in Polytene Chromosomes
by GEORGE T. RUDKIN

The Control of Puffing in Drosophila hydei
by HANS D. BERENDES

Balbiani Ring Activities in Acricotopus lucidus
by REINHARD PANITZ

Contributors

ASHBURNER, MICHAEL, Dr., Department of Genetics, University of Cambridge, Cambridge (Great Britain)

BEERMANN, WOLFGANG, Prof. Dr., Max-Planck-Institut für Biologie, Tübingen (Germany)

BERENDES, HANS D., Dr., Department of Genetics, University of Nijmegen, Nijmegen (The Netherlands)

LEZZI, MARKUS, Dr., Zoologisches Institut der ETH, Labor für Entwicklungsbiologie, Zürich (Switzerland)

PANITZ, REINHARD, Dr., Zentralinstitut für Genetik und Kulturpflanzenforschung der Deutschen Akademie der Wissenschaften, Gatersleben (DDR)

PELLING, CLAUS, Dr., Max-Planck-Institut für Biologie, Tübingen (Germany)

RIBBERT, DIETER, Dr., Zoologisches Institut der Universität Münster, Münster (Germany)

ROBERT, MICHEL, Dr., Institut für Genetik der Universität des Saarlandes, Saarbrücken (Germany)

RUDKIN, GEORGE T., Dr., The Institute for Cancer Research, Philadelphia, Pennsylvania (USA)

Chromomeres and Genes

Wolfgang Beermann

Max-Planck-Institut für Biologie, Tübingen

I. Introduction

The giant cross-banded chromatin "threads" found in the nuclei of various Dipteran cell types had caught the imagination of many cytologists even before their chromosomal nature was established almost forty years ago (Heitz and Bauer, 1933; Painter, 1933; King and Beams, 1934). The revelation that these structures simply presented an enormously magnified but faithful picture of the linear subdivision of the mitotic interphase chromosomes suddenly changed mere curiosity into enthusiasm: In the words of Painter (1934) "it was clear that we had within our grasp the material of which everyone had been dreaming. It was clear . . . that the highway led to the lair of the gene." We must admit today that the goal defined in these prophetic words has not yet been reached, and we cannot even claim that the highway towards this goal has been found, in spite of the concerted efforts of two generations of cytologists, cytochemists, and geneticists to localize and characterize genes in polytene chromosomes. This is not the fault of the material itself, nor is it due to lack of skill on the part of the investigators. Progress has been blocked both by problems of methodology and by the difficulty in defining the basic questions of chromosomal and genetic subunits in physical-chemical terms. Adequate micromethods to study the composition and activity of individual genetic units in polytene chromosomes are only now beginning to become available. On the other hand, it appears increasingly clear that our concepts of chromosome structure and function in the past were oversimplified: On the cytological side the classical, static view of polytene chromosome structure will have to be reevaluated and partly revised. On the molecular level, the tendency of explaining all aspects of the genetic machinery in eukaryotes in terms of simple molecular models derived from the study of prokaryotes will have to be abolished, especially in view of the functional behavior of eukaryotic chromosomes during development and differentiation. Conversely, for an understanding of the developmental processes themselves, not only in the Diptera, it appears imperative to understand the basic aspects of eukaryote chromosomal organisation in terms of structural, functional, and genetic units. The only material in which such a combined approach seems to be feasible is the giant polytene chromosomes of the Diptera.

II. The Chromomeric Subdivision of Polytene Chromosomes

A. Morphological and Quantitative Aspects

The single most distinctive feature of giant polytene chromosome structure, and the one of greatest relevance to genetic and developmental questions, is their banding pattern. Generally speaking, this pattern can be described cytologically as an aperiodic, non-repetitive sequence of band or disc-like elements of high staining intensity and mass concentration (bands or chromomeres) connected by sections of extremely low mass concentration and staining (interbands or interchromomeres). As a rule the combined euchromatic portions of the genome in various Dipteran species display a total of at least 2000 and maximally 5000 bands and interbands. These values are for fully extended chromosomes at polytenic levels which permit complete resolution of all components of the banding pattern. The finding of lower band numbers per genome or per individual chromosome can usually be attributed to incomplete extension of the chromatids during polytenic growth, often concomitant with inaccurate lateral association of the chromatids. This makes resolution of all but the heavier bands difficult, as is the case in the polytene chromosomes of some Dipteran species. In others, this restriction applies only to some cell types or developmental stages (see Beermann, 1962). Although no detailed comparisons have been made, it seems that the number of bands that can ultimately be resolved per unit length of euchromatic genome is not widely divergent in different Dipteran species and families. This statement also holds true for other general characteristics of the banding pattern. The findings on *Drosophila melanogaster* and on *Chironomus tentans*, which will form the basis of the following discussion, can therefore be considered as representative of the conditions encountered in most Dipteran flies.

The morphology of the banding and the bands in polytene chromosomes has been studied almost exclusively in fixed material, either by light or electron microscopy. There is no doubt, however, that essentially the same banding exists in living cells. Fixation and further handling of the chromosomes may introduce artifacts either by changing the structure of the individual band, or by grossly distorting and obscuring the spatial relationships between the bands (cf. Lezzi and Robert, this volume). The latter effect will cause an unexperienced observer to miss some bands and see others twice. Alterations of individual band structure are less easily recognised as artifacts. As a common example we may mention the "capsules" which originate from the vacuolisation of individual heavy bands and which often have been interpreted as "doublet" bands. This artifact can be avoided by quick fixation of the tissue without prior incubation in Ringer solution. Apart from this difficulty, the fixation and staining procedure which reveals and defines the greatest number of bands must be considered the best one, even though its effect may depend on a slight artificial condensation of the band structure. With the introduction of electron-microscopical mapping (Sorsa and Sorsa, 1967; Berendes, 1970), most uncertainties with respect to the number and arrangement of bands in polytene chromosomes have now become resolvable. Earlier electron-microscopic work on sectioned material (Beermann and Bahr, 1954; Lowman, 1956) had already revealed the presence of the same diversity in band thickness, fine structure, and mass condensation that can be seen in the light microscope, although a variety of fixation procedures had been used. The new combined squashing and sectioning technique, as developed by Sorsa and Sorsa, greatly simplifies the location of specific chromosome regions and thus allows the resolving power of the electron microscope to be used for a complete rechecking of published chromosome maps as well as a detailed reevaluation of earlier conclusions with respect to band dimensions and fine structure based on light microscopy.

An important general aspect of band structure, confirmed and emphasized by electron microscopy, is the sharp delimitation between bands and interbands. Apart from being a necessary prerequisite for precise localization, this fact must have significant implications with respect to chromosome structure. Exceptions, i.e., individual bands with diffuse contours or stretches of blurred banding (e.g., 3B1—2 in Fig. 3), can be attributed to puffing (cf. article by Ashburner for further illustrations). The banding may also be ill-defined in the vicinity of "weak spots", i.e., regions containing late replicating material (Arcos-Teran and Beermann, 1968), and in regions with an inherently low accuracy of lateral association of the chromatids (e.g., the so-called "β"-Heterochromatin, Heitz, 1934). The second point of general interest is the question of doublets and multiple bands. Apart from the artifact already mentioned, which makes single bands appear as doublets, it is clear that pairs or triplets of adjacent bands may appear as one, whenever the interbands separating them are very short, so that optical or actual physical fusion occurs. This difficulty should be overcome by studying well-stretched chromosomes in exact longitudinal sections at several planes in the electron microscope. Electron-microscopical studies, as done by Sorsa and Sorsa (1967—1969), as well as by Berendes (1970), show that in a few instances, e.g., 3A3 and 3A4 in the X chromosome of *Drosophila melanogaster*, apparent single bands can be resolved in two constituent half bands of equal thickness (Fig. 3). In others, such as 1B3—4, which have been considered as doublets by Bridges (1938), a split can at least sometimes be recognized, but in the majority of the so-called doublets, no evidence in favor of a subdivision of these bands could be found, a point particularly stressed in the maps published by Berendes (1970). In other words, electron microscopy supports the contention that most bands are of unit character regardless of their thickness. A split as thin as, or even thinner than, 0.05 μ traversing a thick band, would still be recognizable in the electron microscope, as demonstrated in the case of 3A4. Interbands of such small dimensions are, however, not very frequent anyway (see below). A third morphological peculiarity of general interest concerns the state of structural cohesion within individual bands as viewed across the chromosome. Some bands almost always appear as continuous lines of even thickness at all optical planes or all planes of sectioning. Others constantly form a row of equidistant dots or dashes (Fig. 1). Apart from hinting at a chemical individuality of single bands, this behavior emphasizes the difficulty of estimating the relative DNA content of a single band from band thickness alone. A dotted band, if smoothed out in a continuous line, would probably only be half as thick or would contract into a disc of only half its previous cross-sectional area. A fourth and last aspect of band morphology raises a related question: The distribution of dry mass or mass density in individual bands. Although in comparison to interbands all bands obviously show a much higher concentration of dry mass (according to Engström and Ruch, 1951, by a factor of 10 on average), electron-microscopical inspection leaves little doubt that the density of different bands differs appreciably. It seems that the highest density values occur more frequently in thick bands than in thin ones and, in addition, in bands which by other criteria can be considered as heteropycnotic.

Although structural characters, such as granulation and mass concentration, help in identifying bands and stress their individuality, the most relevant character of the elements of the banding pattern, both in theoretical and practical respects, is their

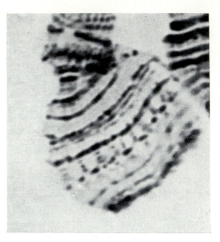

Fig. 1. Granular (dotted) and smooth bands in the flattened-out terminal section of chromosome III from the Malpighian tubules of *Chironomus tentans*. Lactic-aceto-orcein; phase contrast (From Beermann, 1952)

relative thickness. Generally speaking, the thickness of individual bands in typical polytene chromosomes (e.g., from salivary glands of fullgrown larvae of *Drosophila*, *Chironomus*, or *Sciara*) ranges between a maximum of 0.3 μ down to just below the limit of light optical resolution at $0.1-0.05$ μ. Interbands apparently do not exceed 0.2 μ in thickness but otherwise seem to vary similarly. It is clear that band thickness as a parameter depends on many non-specific variables, such as fixation, accidental distortion due to the squashing procedure, or, in electron microscopy, on the angle of sectioning, and, in light microscopy, on the conditions of illumination and other optical parameters. Thus, for instance in Bridges' (1938) map of the X chromosome of *Drosophila melanogaster* band 3A1—2 is represented as being approximately 0.4 microns thick. In electron micrographs (see Figs. 3 and 4), the same band has a thickness of 0.25 μ. Band 3C1 in Bridges' map appears to be 0.2 to 0.25 μ thick, whereas in electron micrographs its thickness is just about half this value. The exaggeration of longitudinal dimensions in Bridges' map of the X chromosome is still more pronounced in the case of the interbands which frequently attain a thickness in excess of 0.5 μ. This is clearly the result of overstretching, since the total length of Bridges' X chromosome is given as 414 μ as opposed to about 200 μ in ordinary preparations and in Bridges' own original map of the X chromosome (1935). In light microscopy, overstretching of a polytene chromosome will not only grossly exaggerate interband distances but also make the bands themselves appear heavier (Fig. 2). In addition, the dimensions of very thin bands preclude a reliable optical delimitation and will thus invariably cause an overestimate of their thickness. It seems worthwhile, therefore, to illustrate the situation by a few recent results obtained with electron microscopy in chromosome regions which had previously been intensively studied by light microscopy.

The tip section of the X chromosome of *Drosophila melanogaster* from 1A1 up to 3C1 inclusive, measures 36 μ in BRIDGES' (1938) map. In a well-extended X chromo-

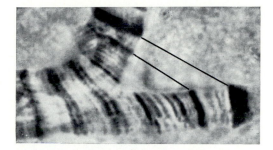

Fig. 2. Differential stretching of two homologous unpaired sections of a salivary gland chromosome of *Chironomus tentans*. Stretching brings out more details and emphasizes the differences between bands. Acetocarmine squash preparation; phase contrast (From BEERMANN, 1952)

some which, after squashing was postfixed for electron microscopy according to SORSA and SORSA (and a portion of which is shown in Fig. 3) the same region measures 23 μ. In another instance, 18 μ were measured under the same conditions. BERENDES (1970) who uses glutaraldehyde fixation prior to squashing in acetic acid, presents a longitudinal section of an X chromosome in which the distance from 1A1 to 3C1 inclusive is 17 μ. The band number for the same region is 102 in BRIDGES' (1938) map, including 27 doublets. BERENDES, on the other hand, counts 71 bands, or granular zones, including only four doublets. This is close to BRIDGES' (1935) original figure of 68 bands, which included only 9 doublets. Expressed otherwise, the band density in terms of band number per micron is 2.8 in BRIDGES' map, and if doublets are counted as one, it is 2.0. In the electron microscope, as well as in BRIDGES' original map, it is between 3.2 and 3.8. In light-microscopical mapping of polytene chromosomes in general, a resolution of 3 bands per micron will only rarely be achieved. In other words, it is necessary to overstretch to a point where only two bands per micron are left to be resolved. Electron microscopy, on the other hand, shows that this procedure is legitimate as long as only the presence or absence of a band are at stake: Stretching does not artificially increase the number of bands per unit length of euchromatic genome. At this point, it is necessary to illustrate the situation by describing the actual sequence and dimensions of bands and interbands in a few selected examples of well-defined chromosome regions as studied in the electron microscope.

Within the tip-section of the X chromosome of *Drosophila melanogaster* a segment of particular genetic interest is the one limited by the genes *zeste* and *white*, or in terms of bands, 3A3 and 3C2. Since there is some uncertainty as to whether the gene white is actually associated with band 3C2, the following description does not include this band. In the electron micrographs, shown in Fig. 3 and 4, this segment

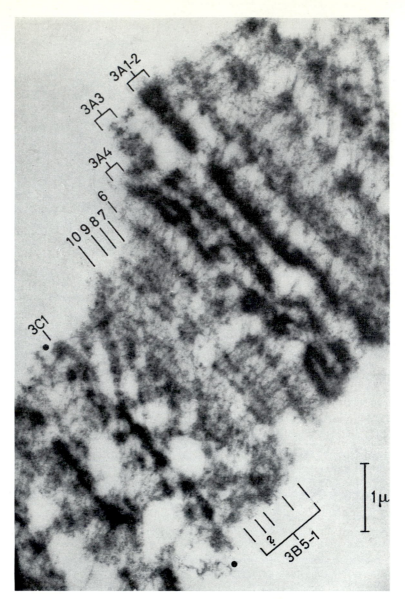

Fig. 3. Banding in the *zeste-white* interval of the *X* chromosome of *Drosophila melanogaster*, as revealed by electron microscopy. This chromosome carries the $zeste^{+\,64\,b9}$ inversion, the distal limit of which is at the dotted band next to 3C1 (see text for further details). Aceto-methanol fixation, uranylacetate staining (Courtesy of Dr. V. Sorsa)

(3A4 to 3C1 inclusive) measures approximately 2.7 microns and contains 12, or if 3A4 is counted as two separate bands, 13 more or less well definable bands. The banding situation within and adjacent to the puffed region 3B1—2 appears somewhat obscure, and possibly another band will have to be added to section 3B in addition to band 3B5, a band which is also not present in Bridges' original map.

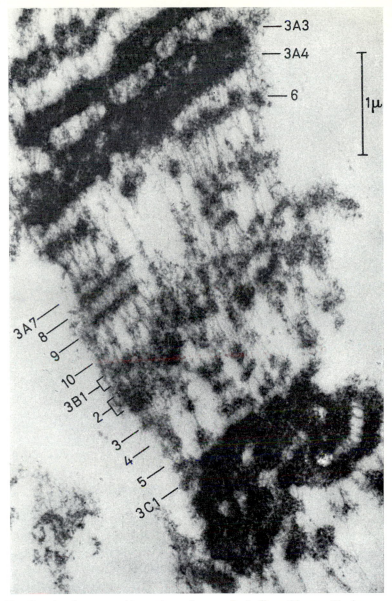

Fig. 4. Banding in the *zeste-white* region of a cytologically wild type X chromosome of *Drosophila melanogaster*, stock w^{66l34}, as revealed by electron microscopy. Technique as in Fig. 3 (Courtesy of Dr. V. Sorsa)

Although the quantitative interpretation of banding, as presented in Table 1, is therefore still tentative in some respects, the following general features of banding of polytene chromosomes are stressed by this example:

1) The thickness both of bands and interbands in chromosomes that are not overstretched ranges over one order of magnitude.

Table 1. Band and interband thicknesses in the zeste-white interval of the X chromosome of *D. melanogaster* as estimated from 2 electron micrographs by Sorsa (cf. Fig. 3 and 4). All values in microns

| | Sorsa 1 (Fig. 3) | | Sorsa 2 (Fig. 4) | |
	Band	Interband	Band	Interband
		0.12		0.07
3A4a	0.12			
		0.05	0.25	
3A4b	0.12			
		0.25		0.15
3A6	0.10		0.12	
		0.20		0.25
3A7	0.03		0.05	
		0.05		0.05
3A8	0.06		0.08	
		0.10		0.10
3A9	0.06		0.08	
		0.15		0.15
3A10	0.03		0.04	
		0.10		0.06
3B1			0.08	
	0.35			0.08
3B2			0.13	
		0.08		0.15
3B3			0.08	
				0.08
3B4	0.60		0.06	
				0.10
3B5			0.06	
				0.10
3C1	0.12		0.12	
Total[a]	1.16	1.63	1.02	1.47

[a] Estimated under the assumption of equal thickness of bands and interbands in those regions where individual measurements were impossible.

2) Apart from the occasional occurrence of twin bands of equal dimensions and texture, such as the two components of 3A4, or bands 3A8 and 3A9, there is no recognizable regularity in the ordering of bands and interbands as related to their thickness. Relatively long interbands may occur adjacent to thin as well as thick bands, and vice versa.

3) Although at least half of the bands in the section shown in Figs. 3 and 4 would have to be classified as being "submicroscopic", they can nevertheless be detected by light microscopy in well extended chromosomes, as shown by the fact that with the exception of one, they have been included by Bridges in his map with approximately correct interband distances as well as correct relative thicknesses. Amazingly, this is even true for bands such as 3A7 and 3A10, whose thickness in the section shown ranges below 0.05 μ. As shown in Fig. 5 and 6, these conclusions are

basically confirmed by an analysis of other chromosome regions in *Drosophila melanogaster* as well as in *Chironomus*.

In order to understand the banding pattern in terms of genetic units, the question of the DNA content of bands and interbands and its possible relation to the parameters just discussed is of crucial interest. With respect to the bands, the situation is relatively simple: If the bands are taken as units and if the concentration of DNA as a first approximation is taken to be the same in all bands, then the relative amount of DNA per band and per chromatid should vary over at least one order of magnitude, as has in fact been found (RUDKIN, CORLETTE, and SCHULTZ, 1956). Either by direct reference to a known DNA standard, or from determining the amount of DNA per haploid genome and dividing by the number of bands, one arrives in the case of *Drosophila* (RUDKIN, 1961; MULDER et al., 1968), or *Chironomus* (EDSTRÖM, 1964) at an average of about 5×10^{-17} g DNA per chromomere per chromatid, corresponding to about 40000 nucleotide pairs. Since, as shown by RUDKIN (1961), and as illustrated by BRIDGES' maps as well as the examples from the quoted studies of SORSA and SORSA and BERENDES, the heavier bands are less frequent than the lighter ones, one can estimate the DNA content of the heaviest bands as being in the vicinity of 100000 base pairs per chromatid, and that of the light ones as ranging around a value of 10000 base pairs per chromatid, with the finest ones possibly being well below 5000 base pairs per chromatid. In these calculations the DNA content of the interbands has been neglected. This is justified because estimates both of dry mass and of DNA content in interbands seem to indicate that the contribution of interband material to the total chromosome mass or DNA cannot exceed 5%. Since the question of interbands is of considerable interest with respect to gene localization, its present status is briefly discussed.

Interbands, by definition, are those sections of a chromatid which connect or separate adjacent bands. This definition depends on the recognisability of bands, just as definition of the bands is dependent on the recognisability of interbands. As is easily seen, the unequal concentration of dry mass in the two types of units, bands and interbands, makes the practical application of this type of definition more of a problem in the case of the interbands than that of the bands. Whereas the presence of a very short interband section within an apparently uniform band can often be revealed by overstretching (see Fig. 3), the existence of a delicate submicroscopic band within a supposedly uniform interband region, is readily obscured by mechanical distortion. All light-microscopic work on so-called "interbands" is subject to this uncertainty. Determinations of dry mass (ENGSTRÖM and RUCH, 1951) as well as of DNA content (SWIFT, 1962) in interbands of *Chironomus* and *Sciara* must therefore be considered as overestimates. In electron microscopy, as the examples selected from the work of SORSA and SORSA and BERENDES show, the length of interband sections ranges from 0.03 to about 0.2 μ, with an average of 0.12 μ per interband. However, closer inspection of electron micrographs casts doubt on the existence of non-banded sections which are more than 0.1 micron in length: As SORSA and SORSA have pointed out repeatedly, longer interband sections often appear to be subdivided by "rows of parallel individual particles or granular chromomeres, the dimensions of which vary from 100–250 Å" (SORSA, 1969; cf. Fig. 5). Unless, as SORSA and SORSA suggest, these particles represent special elements, e.g., initiation sites for replication and not necessarily condensations of DNA,

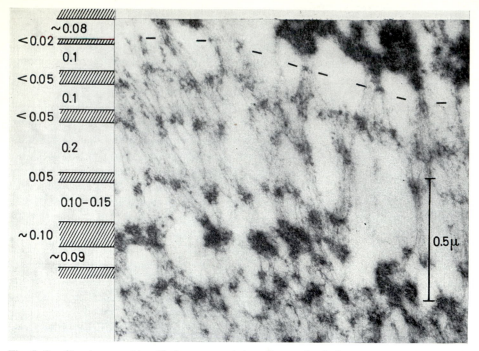

Fig. 5 Banding in an unidentified segment of the salivary gland chromosomes of *Drosophila melanogaster*, as revealed by high resolution electron microscopy. Thickness estimates of bands and interbands are given in microns (Adapted from Sorsa and Sorsa, 1968)

one would have to conclude that DNA bands finer than those previously thought to be the smallest class may occur in giant chromosomes.

In light as well as electron micrographs the interband regions show a typical longitudinal striation which has been interpreted as a visible manifestation of the state of polyteny. In electron micrographs the number of definable fibrils per cross section seems to approach the calculated number of chromatids so that the thinnest fibers may represent individual chromatids. Sorsa and Sorsa (1968) find that the interband fibrils are about 120—170 Å wide and that they often seem to be composed of two smaller fibrillar sub-units of 50 to 60 Å in width. Among other things, these observations seem to imply that there is no close lateral association of individual chromatids in the interband sections which leaves the bands as the only sites of specific attraction between homologous chromosomes. If an interband fibril about 150 Å wide represents a single chromatid, then the chromatid cannot include more than two DNA double helices side by side. In other words, we may either be dealing with two almost extended daughter double helices or with one moderately supercoiled double helix (plus the attached histones; cf. Ris and Kubai, 1970). The average interband with an estimated length of 0.1 μ (this may actually be a maximum estimate) can thus hardly be equivalent to much more than 0.5 μ of extended DNA α-helix, corresponding to a maximum of 2000 nucleotide pairs. The possibility, which is also inherent in the results of Swift (1962), that the DNA in interband fibers is relatively

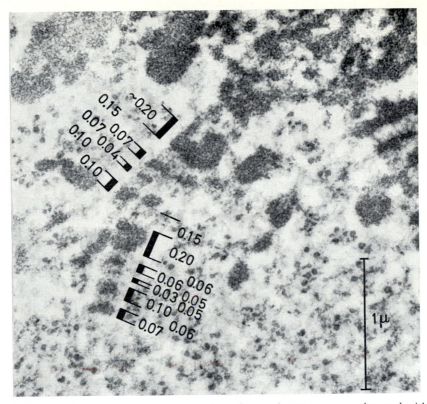

Fig. 6. Banding in two very fine branches of a polytene chromosome, as observed within a Balbiani ring in salivary chromosome IV of *Chironomus tentans*. Formaline-OsO$_4$ fixation, uranyl acetate staining; thin section of unsquashed salivary gland nucleus. Estimated thicknesses of bands and interbands are given in microns. Original

extended, should be checked using modern methods of electron-microscopical densitometry (BAHR, 1970), because, as will be pointed out later in this article, the maximum length of DNA that can be accommodated in an interband is of critical interest to the question of gene localisation in the polytene chromosome.

Using all the available data (cf. extensive review by RUDKIN, this volume) and keeping the necessary reservations in mind, the following picture of the distribution of DNA in bands and interbands along a typical polytene chromosome, such as the X chromosome of *Drosophila melanogaster* emerges (Table 2): Although at least half of the length of the chromosome must be in the interband sections, more than 95% of the DNA is found concentrated in the bands. The total DNA per chromatid, as calculated from the data in Table 2 would amount to 3.0×10^{-2} pg for the euchromatic portion of the X chromosome. This corresponds rather well with the results of RASCH (1971), who found approximately 0.2 pg of DNA per haploid genome in the sperm head of *Drosophila melanogaster*. One fifth of this value, i.e., 0.04 pg, would roughly correspond to the DNA content of the X chromosome. Since 3.5×10^{-2} pg of DNA correspond to approximately 10 mm of extended

Table 2. Estimated cytological and DNA length in chromomeres and interchromomeres of the euchromatic portion of the salivary X chromosome of *Drosophila melanogaster*.

	Number[a]	Total cytolog. length[b]	Individual thickness (average)[b]	Individual DNA content (average)	total DNA length	sum DNA/chromatid X euchromatin
Bands	800	100 μ	0.12 μ	10 μ = 30000 b.p.	8000 μ = 95%	~3.0 × 10⁻² pg
Interbands	800	100 μ	0.12 μ	<0.5 μ = 2000 b.p.	400 μ = 5%	

[a] rounded figures, counting doublets as one.
[b] estimated extrapolation from combined data of Figs. 3—6, and Table 1.

double helix, the ratio of DNA length to polytene chromosome length (0.2 mm) is of the order of 50 to 1. This means that in the bands the same ratio must be close to 100 to 1, whereas in the interbands it should be less than 5 to 1.

B. Constancy of the Banding Pattern

Strictly speaking, the banding pattern of a polytene chromosome must be considered as part of the phenotype, inasmuch as the differential condensation of DNA along the chromosome axis is almost certainly the result of an interaction of the DNA with different proteins, such as histones. Nevertheless, in view of the excellent precision with which bands and interbands are delimited and follow one another in a characteristic sequence, the banding pattern has often been considered as an invariant structural attribute of the genetic material itself. However, the fact that in a given chromosome section the banding is precisely the same in all cells of a given tissue at a given developmental stage, such as salivary glands of 3rd instar larvae, only shows that at the level of the DNA backbone of each chromatid some kind of linear differentiation, or subdivision, must exist which, in a given developmental situation, always expresses itself in the same way. A careful comparative study of the banding pattern of polytene chromosomes in different tissues of *Chironomus* has indeed revealed subtle tissue-specific variations in the microscopic appearance of single bands, or small groups of bands. As described in more detail in Beermann (1952) and Beermann (1962), the extent of this variation is rather small and in many cases due to differential puffing. In other words, the general characteristics of banding, such as number of bands per unit length as well as the specific banding sequence of homologous chromosome sections, are nearly identical. Changes which are probably not due to differential puffing can be classified as follows:

1) Presence—absence: A specific band that is clearly seen in one or several different tissues cannot be demonstrated in another tissue. This has only been found in the case of very thin bands.

2) Singlet-doublet: A band may always appear double in one tissue and always as a unit in another.

3) Differential spacing: The distance relationships within a group of bands (3 or more) appear specifically changed in one tissue as compared to the others. Changes of this type, as well as those of type 2, might reflect tissue specific differences

in the length, or state of condensation, of interbands. Another typical variant could likewise be due to differential interband behavior:

4) Tissue-specific increased staining of the interbands connecting two or more adjacent heavy bands. The increase might indicate true changes in the DNA content of the interbands concerned, in analogy to the well-known DNA puffs in *Sciara* which seem to originate in the interval between two closely adjacent bands (BEERMANN, unpublished).

5) Differences in the intensity of staining of bands: These could again reflect tissue-specific changes in DNA content. As long as the existence of such DNA changes has not been directly demonstrated, it appears safer to explain the staining differences by differential condensation, i.e., puffing.

None of the limited data on banding differences that are at present available are in conflict with one important generalization: The number and sequence of bands and interbands are constant in all tested somatic tissues. All apparent changes in the banding pattern, including those which are possibly not due to differential puffing, can be attributed, at least hypothetically, to changes within individual bands or interbands. There is no indication for the existence of specific differences in the pattern of chromosomal subdivision, i.e., band individuality, as such. A given band or interband may look specifically different in different cell types, but nevertheless it seems to represent the same individual subunit or subsection of the genome.

III. Methods and Results of Gene Localization in Polytene Chromosomes

A. Numerical Relations between Bands and Genes

The two most commonly used notions of a gene in terms of molecular genetics are those of the "cistron" and the "operon". Theoretically, the cistron is defined as a nucleotide sequence coding for one polypeptide, whereas an operon should comprise two or more adjacent cistrons which are transcribed as a unit. Since the genetic analysis in eukaryotic organisms can only rarely be reduced to the level of single polypeptides, the experimental distinction between cistrons on the one hand and operon-like units on the other, mainly rests on the classical test of allelism, or complementation. Mutations, visible or lethal, which when compounded in the heterozygote do not complement each other, i.e., do not manifest the wild type condition, are said to be allelic and belong to the same genetic unit. If two adjacent mutants are found, which do complement each other, they can as a rule be considered as belonging to different cistrons. The question as to whether or not such adjacent cistrons are members of a higher unit of the operon type should theoretically be resolvable by searching for point mutants which do not complement either of the two adjacent complementing mutations ($0°$ or polar mutations). On recombination analysis, such mutants should all map towards one end of the presumed operon, thus defining the starting point of transcription and translation. The situation could be somewhat complicated by the occurrence of complementation between mutants within the same cistron (intra-allelic complementation), inasmuch as this would make a single cistron sometimes appear as a multicistronic unit. On the other hand, allelic polar mutations within one operon may be indistinguishable on the phenotypic level from allelic point mutants within a single cistron, thus decreasing the apparent number of cistrons. In determining the actual number of cistrons in eukaryotic chromosomes, it is necessary to "saturate" a given chromosomal section with mutants to such an extent that operator-type mutations (polar mutants) can be recognized as such. The same applies to the recognition of complementing intra-cistronic mutants.

Early attempts to estimate the number of "genes" in a given chromosome or genome were simply based on the assumption that the rate of induced mutations of all functional units, as defined by allelism, is similar. Thus, by determining the lethal mutation rate at a given locus and comparing it with the lethal mutation rate of the entire chromosome, Alikhanian (1937) arrived at the conclusion that the X chromosome of *Drosophila melanogaster* contains 968 genes. At the time when these findings were made they were not considered as surprising, since they agreed with other estimates of gene number and with the widely held view that the bands of giant polytene chromosomes represented the genes. From a modern point of view, however, the same result would appear highly problematic in view of the DNA content of the X chromosome of *Drosophila melanogaster*, which theoretically allows for the presence of at least ten times as many "genes".

The early estimate made by Alikhanian could be dismissed as being subject to all kinds of errors, were it not for the agreement, already mentioned, with the total band number of the X chromosome. The proposition that an inherent 1 : 1 relationship between bands and genes might exist has recently been checked again, using the saturation of defined chromosome regions with induced lethal mutations: By far the most extensive work of this type has been carried out by Judd, Shen, and Kaufman (1972). They have studied the genetic fine structure of a segment of the X chromosome of *Drosophila melanogaster*, which is defined by a deficiency originally obtained by Judd (1961), $Df(1) w^{rJ1}$. This deficiency includes bands 3A2 through 3C2, i.e., the same section whose banding has been described in detail in the foregoing chapter and in Figs. 3 and 4. Genetically, it extends from *zeste* to *white* inclusive.[1] By means of X rays and chemical mutagens such as EMS, ICR-170, ethyleneimine, and N-methyl-N¹-nitro-N-Nitrosoguanidine, Judd and associates have induced a total of 116 lethal or semi-lethal mutations that map between z and w. By complementation analysis, including the use of overlapping deletions and duplications, these mutants could be assigned to not more than twelve functional units between the genes *zeste* and *white* (Fig. 7). Although the number of mutations found per unit varies considerably (between 2 and 34) and does not follow a Poisson distribution, the fact that at least 5 alleles have been found in 9 out of 12 of these units would seem to preclude the possibility that many of them could be multi-cistronic "operons", or that many additional units exist in the tested segment. Thus, it is hard to escape the conclusion that the section delimited by *zeste* and *white* does not contain more than 12 apparent cistrons. This number is exactly the same as that of the bands counted between 3A3 and 3C2 in Bridges' (1938) as well as in Berendes' (1970) map based on electron microscopy. In the latter, 3A4 is counted as two bands and section 3B shows a fifth band whereas bands 3A5 and 3A10 are omitted. Judd (l.c.) bases his own interpretation of the cytological situation situation on Berendes' map with the exception of 3A4 which he considers as unitary. The crowding of comple-

---

1 Recently it has been found that $Df(1)w^{rJ1}$ is also deficient for the gene *tko* which is immediately to the left of zeste and thus seems to be associated with 3A2 (Judd et al., 1971).

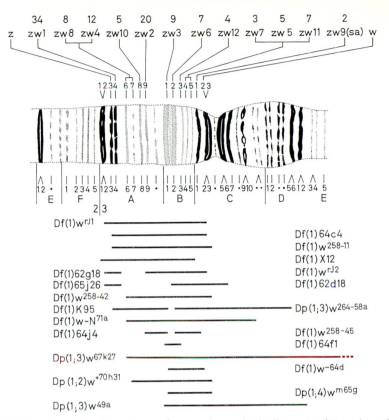

Fig. 7. Co-linearity of bands and complementation units in the *zeste-white* region of the X chromosome of *Drosophila melanogaster*, as demonstrated by JUDD. Cytological map of BRIDGES (1938) slightly modified according to EM results (see Figs. 3 and 4). *z zeste*: *w white*; *sa sparse arista. zw* 1—12: lethal, or semilathal loci between *z* and *w*, as defined by complementation. Number of allelic mutants found at each locus are given on top. *Df, Dp*: Deficiencies and duplications used in the ordering and localization of complementation units (Adapted from JUDD et al., 1971, Figs. 2 and 5)

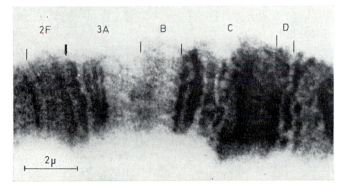

Fig. 8. Electron micrograph of a thick section of the tip of the salivary X chromosome from *Drosophila melanogaster*, showing many fine bands in region 3B. Glutaraldehyde fixation, acetic acid squash. (From BERENDES, 1970)

mentation units in subdivision 3B, as indicated by deletion mapping (cf. Fig. 7), corresponds to the crowding of very fine bands which is occasionally observed in EM pictures of 3B and which indicates the presence of at least one further band there (Fig. 8). On the other hand, in Sorsa's EM photographs (Figs. 3 and 4) band 3A10 can clearly be seen, a band which according to Judd's present interpretation would have no function at all. However, although some of the details of the cytology of Judd's segment still need clarification, the data nevertheless strongly suggest a 1 : 1 relationship between complementation units (apparent cistrons) and bands. It is interesting that a recently discovered non-lethal "clock mutant" has also been found to be situated in this segment (Konopka and Benzer, 1971).

Lifschytz and Falk (1968; 1969) have performed a similar study in another segment of the X chromosome of *Drosophila melanogaster* which is, however, less well-definable cytologically. They carried out allelism tests with 35 X ray induced and 70 chemically induced lethals in the region covered by the $Y ma-l^+$ translocation. For this segment, a consistent linear map of 35 complementation units could be constructed. The distribution of mutations over the segment was highly non-random. Thus, even though an average of 2 point mutations per functional unit was obtained, the results cannot yet be considered as conclusive. On the other hand, there seems to be little probability that many more units will be discovered in further mutation tests .According to Schalet et al. (1970) the chromosomal segment studied by Falk and co-workers extends from a point within 18F through sections 19 and 20 in Bridges' map. This corresponds to 32 bands up to section 20A. In addition, the $Y ma-l^+$ chromosome should contain the bands 1A1 through 1A4 which include the locus of *yellow*. It follows that again the number of functional units as defined by complementation analysis is very close, if not identical, to that of the number of bands in this segment. According to Falk (personal communication), a 1 : 1 relationship between bands and essential functional units has also been found to exist in a segment including the gene *prune* near the left end of the X chromosome of *Drosophila melanogaster*.

Still another set of data indicating the exact agreement between numbers of polytene chromosome bands and numbers of essential "genes" has recently been presented by Hochman (1971). In an attempt to determine the gene content of the small 4th chromosome of *Drosophila melanogaster*, Hochman carried out allelism tests involving some 100 chemically induced mutations and over 70 previously detected lethals and visibles. The 170 mutants were found to belong to 40 separable genetic units, 33 of which were "vital" ones and 7 recognized through visible mutant effects. The frequency of repeated mutations in individual loci indicates that "saturation" has almost been achieved, in other words there can hardly be more than 50 essential loci on chromosome IV. The number of bands in chromosome IV according to Bridges (1942) likewise approximates 50, if "doublets" are counted twice. Furthermore, as Hochman points out, if the gene and band numbers are compared in 3 individual regions of chromosome IV, the agreement is again excellent.

B. Delimitation of Individual Loci

The fact that, as described in the preceding paragraph, the number of essential "genes" or functional units in a given chromosome segment is always in close

agreement with the number of crossbands of the same segment as observed in the
polytene state, suggests but does not prove that for each essential functional unit
there is just one band and/or one interband. Several attempts have been made to
prove this relationship directly, using classical cytogenetic localization techniques.
Since the principles of these techniques are well known, they will not be described

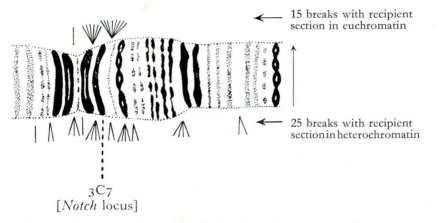

15 breaks with recipient
section in euchromatin

25 breaks with recipient
section in heterochromatin

3C7
[*Notch* locus]

Fig. 9. Cytological break points of radiation induced chromosomal rearrangements
associated with a *Notch* phenotype (From DEMEREC, 1941)

here in detail with the exception of those aspects which have bearing on the central
question itself. In order to determine the approximate position of a known genetic
function in a giant polytene chromosome, it is sufficient as MACKENSEN (1935)
and others have shown, to collect data on the break points of chromosomal aberrations which were induced along with X ray induced mutations of the function in
question. X ray induced visible or lethal mutations are in fact often associated with
rearrangements such as inversions and translocations. It is regularly found that
rearrangements associated with mutations of the same functional unit tend to have
one of their break points in common. Unless heterochromatin is involved, this
break point usually falls within the limits of just one salivary gland chromosome
band, as shown for instance by DEMEREC (1939) in the case of *Notch* (see Fig. 9).
It appears that at least in this instance, breaks not only within but also immediately
to the right as well as to the left of the band involved can be associated with inactivation of the unit in question. However, even if the identification of band 3C7
in all of the rearrangements reported were beyond doubt, the theoretical significance
of such a result is hard to assess in view of the relatively gross structural effects which
can be inflicted by X rays. This type of experiment, therefore, does not carry beyond
the statement that a given band (including both adjacent interbands) is most likely
to be the site of the genetic function under study.

Translocations and inversions may also be used in obtaining another more
general type of information on the localization of genes, i.e., the position of a given
mutant with respect to the limits of such rearrangements. In doing so, one could
work with rearrangements which are completely wild type with respect to the

functions under consideration and thus avoid ambiguities as mentioned above. However, in actual practice the chance of discovering a wild type rearrangement with a break immediately adjacent to a given gene is of course extremely rare (but see Fig. 12).

Systematic studies aiming at the question of the localization and physical extension of individual functional genetic units in giant chromosomes or, vice versa, the genetic content of individual bands, are traditionally based on the method of over-lapping deficiencies and duplications (e.g., Slizynska, 1938). A cytologically deficient chromosome which in the heterozygote uncovers a mutant allele of the gene under study must either be structurally deficient for that gene or itself carry a mutant allele of it. The latter alternative can be ruled out, if the cytological extent of several deficiencies involving the same genetic functional unit is compared. True structural deficiencies of a gene should cytologically overlap, or at least be contiguous. Since the smallest cytological unit whose absence can be established by light micro-scopy is usually represented by the single band, it is clear that the shortest overlap detectable by this method must always include at least one band. Non-overlapping, contiguous deficiencies affecting the same genetic function could, if they occurred frequently, be considered as evidence for overlaps entirely confined to interband regions, especially if both adjacent bands were involved with equal frequency. However, such instances seem to be very rare, and if they occur they can also be explained as representing cases where only part of a given band has been removed. The same reasoning applies to duplications. Nevertheless, the study of duplications and deficiencies should at least allow one to answer the question as to whether or not functional genetic units in an organism such as *Drosophila* extend over more than one band. With the possible exception of the complex *bithorax* region (*bx-bxd*, Lewis, 1963) such a situation has indeed never been encountered. A considerable number of visible mutants in *Drosophila melanogaster* have each been found to be associated with just one band or doublet as inferred from the study of overlapping deficiencies and duplications, thus supporting the numerical evidence for a 1 : 1 relationship between bands and genes (cf. Lindsley and Grell, 1968).

That a given "gene" does not, as a rule, extend over a region greater than that defined by an individual band, was somehow to be expected on the basis of the DNA content of bands. The central question raised by the general 1 : 1 relation between bands and genes is, of course, another one: Is there always one, and only one, gene per band, regardless of its DNA content? Only in the case of the *white-Notch*-region in the X chromosome of *Drosophila melanogaster* has the cytogenetic analysis ap-proached the degree of completion required to answer such a question. Although in cytological as well as in genetical terms the situation is far from clear, it may serve to illustrate the methodological problems encountered and at least provide a preliminary answer to our problem. Cytologically, the region comprises bands 3C1 to 3C7 inclusive, and genetically, it contains the loci of the following visible mutants: *sparse arista* (*sa*, Rayle and Green, 1968), white (*w*), *roughest* (*rst*), *verticals* (*vt*), and *facet-split-Notch* (*fa-spl-N*). In accepting for the moment the description and the nomination of banding given in Bridges' map, the classical studies of Demerec, Slizynska, and Sutton (1938—1941, cf. Lindsley and Grell, 1968) soon led to the assignment of *white* to band 3C1 (or possibly 3C2-3), of the gene *roughest* to band 3C4 or 3C5, and of the complex *Notch* locus to band 3C7. In the meantime, considerable efforts have been made to correct or confirm these early localization data, and to fill the gaps left between the loci thus delimited.

The need to reconsider the situation in bands 3C1 to 3 became apparent when it was descovered by GREEN (1959) that a synthetic deficiency, lacking bands 3C2 and 3 but not band 3C1, $Df(1)w^{m4L-rst3R}$. which is male lethal and deficient for the *white* locus (PANSHIN, 1941) is viable in heterozygous combinations with $Df(1)w^{258-45}$ (lacking 3C1) but white-eyed. This could either be explained by assuming that the cytological determinations are incorrect, or by postulating that the *white* locus is situated between bands 3C1 and 3C2. A complex regulatory relationship between bands 3C1 and 3C2-3 has also been envisaged (BEERMANN, 1962). With respect to the cytology, LEFEVRE and WILKINS (1966) as well as GERSH (1967) agree that indeed one of the critical deficiencies, $Df(1)w^{258-45}$, extends into the doublet 3C2-3, thus making 3C2 the most probable candidate for the *white* locus. On the other hand, the reliability of the synthetic deficiency, $Df(1)w^{m4L-rst3R}$, as a tester for the localization of *white* entirely rests on the accuracy with which the left break point of inversion w^{m4} can be determined. According to SUTTON it is between 3C1 and 3C2. The inclusion of at least part of 3C1 in the inversion can, however, not be excluded, since the right hand break point is in the centric heterochromatin of the X chromosome. LEFEVRE and WILKINS (1966) were, however, successful in finding two further *white* deficiencies lacking band 3C2 but not band 3C1, $Df(1)w-ec^{64d}$, deficient for bands 3C2 to 3E7 inclusive, and $Df(1)w^{m4L}-w^{mJR}$, deficient for band 3C2 only

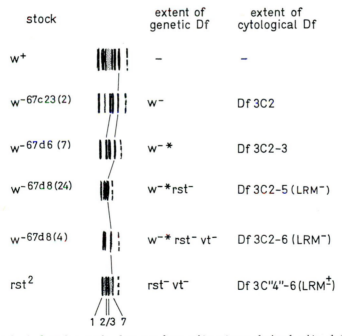

Fig. 10. Cytological and genetic data on four *white-crimson* derived *white* deficiencies, as well as $Df(1)rst^2$. The section shown extends from 3C1 to 3C9-10 inclusive. * lowered vitality, LRM late replicating material (Unpublished work of GREEN, ARCOS-TERAN, and BEERMANN)

and male viable. The first of these is cytologically unambiguaus in showing the presence of 3C1, since the right hand limit of the deletion falls in a region of light and diffuse bands only. In the case of $Df(1)w^{m4L}\text{-}w^{mJR}$, the presence of 3C1 is supported by the viability of this chromosome (see below). This, in turn, would relieve some of the doubts raised against the cytological situation in inversion w^{m4}. The same follows from the finding by Lefevre and Wilkins that a duplication for 3C1 which was obtained by X irradiation from a w^{m4} chromosome complements the lethality of $Df(1)w^{258-45}$ without covering the *white* deficiency itself.

The association of the *white* gene with band 3C2 (rather than 3C1) thus seems to be established. Indeed, Green has recently obtained a series of *white* deficiencies which were fully viable and which, on cytological inspection (Beermann, in preparation), were clearly deficient for the greater portion of the doublet 3C2-3, but not for 3C1 (Figs. 10 and 11). In other words, further essential genetic functions other than those of *white*+ itself have not been impaired by the deletion of band 3C2 (and possibly of a part of 3C3 as well if the two bands are considered as separate units of equal size). This finding, however, still leaves open the question of the actual extension and localization of the *white*+ functions within the deleted portions of 3C2-3. Is all the genetic material of 3C2 "essential" for the function of the *white* gene? As is easily seen, this question is much more difficult to attack experimentally than the mere association of genetic functions with bands. It requires the cytological study of deficiencies immediately to the left or to the right of the *white* gene which do not impair its function. For instance, the gene immediately to the left of *white* on the genetic map is *sparse arista* (see Rayle and Green, 1960). Most probably this gene would be associated with 3C1 and it could be used to produce deficiencies defining the left hand limit of the *white* locus which might be anywhere from the right hand border of 3C1 through well into 3C2. The greater portion of 3C1 may be excluded, if the cytological interpretation of the left break point of inversion *white-mottled* 4 (w^{m4}, Sutton) is accepted as being between 3C1 and 3C2.

More precise data are now available as regards the possible extension of the *white* locus to the right (Fig. 12). $Df(1)rst^2$ which is *white*+, but *rst*− and *vt*−, extends from 3C4 through 3C6 and does not appear to include portions of 3C2-3. Inversion w^{mJ} which does not include the *white* locus (Lefevre, 1952) is said to

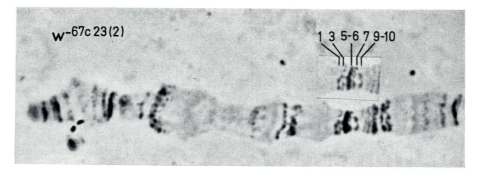

Fig. 11. Exclusive deletion of band 3C2 in a spontaneous, *white-crimson* derived, fully vital *white* deficiency (cf. Figs. 10 and 12). Lactic-acetic-orcein, phase contrast (inset: without phase contrast); chromosome partially overstretched. Appr. 2000×

have its left hand break point between 3C2 and 3C3 and thus seems to exclude "3C3" as the possible site of *white* functions. However, as pointed out before, the localization of inversion breaks adjacent to heterochromatin is doubtful. Lefevre and Wilkins (1966) describe an X ray induced deficiency, $Df(1)N^{63b}$, which seems to come closer to a definition of the right hand border of *white*. Cytologically, this deficiency manifests itself as deletion of at least half of the doublet 3C2-3 and of all subsequent bands up to 3E9. Genetically, it is not deficient for the *white* locus though producing a mutant phenotype similar to the well-known allele *white-spotted* (w^{sp}), which is known to occupy the rightmost recombinational site of the *white* gene. This finding implies that *white* maximally extends about halfway into the doublet 3C2-3, a conclusion which has recently been confirmed and extended through studies on rearrangements showing a changed interaction between *white* and the gene *zeste*:

An inversion, z^{+64b9}, derived by X-irradiation from an X chromosome carrying *zeste* and an intragenic *white* triplication (Green, 1966) has been recovered in which at first sight the left hand break point appears to fall between 3C1 and 3C2 and which nevertheless does not include the functional *white* locus (Fig. 12). In other words, genetic and cytogenetic tests unmistakably show that the *white* gene must be associated with the non-inverted initial part of region 3C which, in this case, seems to

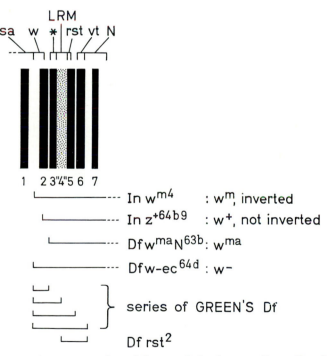

Fig. 12. Diagrammatic representation of the association between "genes" and bands in the *white-Notch* inverval (bands 3C1-7) of the X chromosome of *Drosophila melanogaster*. Only the critical inversions and deficiencies have been included. * vitality factor, LRM late replicating material. w^m (*white*-mottled) and w^{ma} (*white*-marbled) have been considered as equivalent to *white* plus (*white* function still present). For further details see text

be represented by 3C1 only. In well-stretched light-microscopic preparations, a thin faint band immediately to the right of 3C1 can sometimes be seen in the 64b9 chromosome which could either be a remnant of 3C2 or represent band 12B8 from the adjacent inverted portions of region 12. The presence of this band can be demonstrated much better by electron microscopy (Sorsa, unpublished) where it appears in the shape of a dotted band about half the width of 3C1 (cf. Fig. 3). Should this band prove to be part of 3C2 then the *white* gene cannot extend further into 3C2 than by approximately half its original thickness: It remains to be seen whether or not the limits of *white* can still be pushed further back towards the interband between 3C1 and 3C2, or even into a portion of 3C1. At any rate, the essential part of the *white* gene seems to occupy a surprisingly small fraction of the segment called "3C2-3", if it is not exclusively located in the interband between 3C1 and 3C2[2].

If one proceeds further to the right of the *white* gene (cf. Figs. 10 and 12), the next genetic function that can be demonstrated is one whose loss manifests itself as decrease in vitality and which is apparently associated with band "3C3" (e.g., Lefevre and Green, 1971). As mentioned above, however, *white* deficiencies have been recovered by Green in which more than three quarters of doublet 3C2-3 are deleted yet in which vitality is not appreciably diminished. The vitality factor can thus not be associated with more than half (the right-hand half possibly) of "3C3". The conclusion seems unescapable that the majority of the genetic material represented by 3C2-3 is some kind of nonessential "spacer" between *white* and its immediate neighbor to the right. Deficiencies which would precisely delimit the "3C3 factor" on its right side are difficult to obtain for two reasons: (1) Cytological study is hampered by the presence of interstitial heterochromatin (see below) in the gap between 3C3 and 3C5, and (2). Any loss of genetic material between 3C3 and 3C7 is apt to lead to a depression of vitality indistinguishable from that caused by deletion of 3C3 itself. In addition, a complex interaction between the 3C3 factor and another one located in 3C6 has recently been observed (Lefevre and Green, 1971).

Nevertheless, on account of its late replicating behavior the heterochromatic material between 3C3 and 3C5, which must in fact correspond to what in Bridges' map has been represented as "band" 3C4, can be used as a crude marker to determine the most probable limit of the 3C3 factor to the right. *White* deficiencies derived from the mutable *white-crimson* stock (cf. Green, 1967) have been obtained in which the doublet 3C2-3 appears entirely deleted. Two of these deficiencies which show the described loss of vitality do not exhibit loss of late replicating material (LRM) as judged both by the frequency and by the intensity of spot labeling in region 3C after a short tritiated thymidine pulse (Arcos-Teran, in preparation). Thus the vitality factor would remain confined either to band 3C3, or possibly to a small terminal section of this band, or to a minute initial portion of "3C4".

In proceeding further to the right, the next known locus both in recombinational and in cytogenetic terms is that of *roughest* (*rst*). The *rst*+ functions have usually been associated with "3C4", based on the early work of Slinzynska (1938) as well as results obtained with a synthetic deficiency purportedly lacking only this band (Gersh, 1965). In both instances, however, the cytology is inherently unprecise,

---

2 In the case of the *white* gene this conclusion is the more surprising since there are indications that the *white*+ functions depend on the activity of more than one cistron. This situation cannot be discussed here.

i.e., an involvement, at least a partial one, of band 3C5 cannot be excluded. Recently, in the series of mutants derived from *white-crimson*, deficiencies of the late replicating material have been discovered which remain rst^+ whereas two other ones that included *rst* were cytologically deficient not only for all of the LRM but also for 3C5, or speaking more correctly, for approximately one half of the doublet 3C5-6. On the other hand, $Df(1)$ rst^2 which does not include white and which is deficient for bands 3C4 to 6 inclusive (cf. LINDSLEY and GRELL, 1968, and Fig. 10) shows a partial reduction of late replication in region 3C (ARCOS-TERAN, in preparation), indicating that the late replicating material, or "3C4", and the *roughest* locus are closely adjacent to each other but not necessarily identical in their extent. In two synthetic deficiencies of the doublet 3C5-6 which allegedly did not include "3C4" GERSH (1965) found the *roughest* gene still functional (as shown by variegation for rst^+). At least in one of these the cytology is reliable enough to suggest that indeed most of 3C5 has gone in addition to 3C6 so that here is the third locus that does not seem to occupy an entire band, be it "3C4" or "3C5".

In the same study, GERSH discovered a genetic function which phenotypically is clearly separable from *roughest* in that its loss produces a characteristic bristle effect ("*verticals*", *vt*), originally considered as part of the *roughest* phenotype. This effect manifests itself in genotypes which are deficient for the doublet 3C5-6, but not for "3C4" or "3C7". *White-crimson* derived *white* deficiencies which include band 3C5 do not exhibit the *verticals* phenotype which, therefore, must be associated with 3C6. It is interesting, however, that as LEFEVRE and GREEN (1971) have observed, the expression of *verticals* seems to depend on the simultaneous heterozygous absence of genetic material to the left of roughest which they interpret as 3C3 and further that the homozygous absence of the latter in addition to heterozygous deficiency of 3C6 also leads to a *verticals* phenotype. This situation needs further clarification.

The rightmost known locus in region 3C is that of the "complex gene", *split-Notch*. Genetically it comprises an array of recessive and dominant, lethal or non-lethal visible, non-complementing or partially complementing mutants (WELSHONS, 1965) which obviously all belong to the same functional genetic unit. This unit, as mentioned above, has been shown by classical cytogenetic studies to be associated with band 3C7. Specifically the dominant *Notch* phenotype which seems to result whenever the entire unit is deleted or inactivated is frequently found in association with visible structural changes involving band 3C7, including a deficiency of this band only. The actual cytological limits of the *Notch* gene, however, are not as well determined as those of the other genes in the 3C region. The existence of deficiencies of doublet 3C5-6 which do not affect the *Notch* function, makes it fairly clear that the *Notch* gene could not extend much further to the left of 3C7 than the interband between 3C6 and 7. To the right, however, deficiency, or duplication data for the closely adjacent bands 3C8 and 3C9-10 are not available.

The cytogenetic data summarized above show that, at least in the chromosomal segment 3C2 through 3C6, there is no more than one gene per band and that each gene probably occupies only a fraction of the band associated with it (or the adjacent interband). Two or more separable genetic functions within the same band have not been found. The possibility that more than one gene may be associated with a band has been considered with respect to 3C1 in a study of the gene *sparse arista* (*sa*) and lethal factors to the left of it (RAYLE and GREEN, 1968). In recombinational terms the locus of *sa* is 0.1 units to the left of *white*. Since *sa*+ is not affected by deficiency for 3C2 it is probably located in band 3C1. This is supported by the fact that $Df(1)w^{258-45}$ which appears to lack little more than band 3C1 is deficient for *sa*+ in addition to being deficient for *white*+. However, according to Judd et al. (1971), four independent recessive lethal loci are also uncovered by $Df(1)w^{258-45}$. Thus, an apparent one-band deficiency seems to cover 6 independent genetic units. However, as first pointed out by Lefevre, $Df(1)w^{258-45}$ is cytologically deficient not only for 3C1 (and, possibly, a fraction of 3C2, as stated before) but also for the terminal portion of section 3B which in electron microscopy clearly shows the presence of at least 3 (sometimes 4) thin bands in this region (Figs. 7 and 8).

IV. Chromomeres — Units of Gene Regulation in Development?

The relation between chromomeres and genes as revealed by the data available at present, can now be described as follows, at least for the case of *Drosophila*: 1) The number of "essential" functional genetic units in a given chromosome segment is equal to that of the bands or chromomeres in that segment. 2) Only one essential genetic function, or apparent cistron, can be associated with each chromomere or band-interband-complex (a chromomere plus one of its adjacent interchromomeres). 3) The essential portion of each chromomere-interchromomere complex, at least in the case of heavier bands, is restricted to a small "initial" fraction at the edge of the band or, alternatively, only occupies the interband portion.

On the basis of recombination data the latter point has recently been stressed by Lefevre (1971): He presents evidence to show that the mutant sites of the *vermilion* (*v*) gene which is associated with the heavy band 10A1-2 all map in "a short interval about 0.10–0.15 map units from the left edge but nearly 0.5 map units from the right edge of 10A1-2."

It is interesting, that the strange behavior of the chromosomes of some ciliate protozoa during development of the macronucleus also seems to support the concept of the functional nonessentiality of a major portion of the DNA in eukaryotic chromomeres: Ammermann (1970 and earlier) found that in Stylonychia and Euplotes during an initial phase of growth of the macronuclear Anlagen banded polytene chromosomes develop; these then disintegrate so that each chromomere is encapsulated by a membrane followed by degradation and loss of most of its DNA. Only 10 percent of the total nuclear DNA remain. In a second phase of growth the residual DNA increases again by repeated replication. Prescott et al. (1971) found that replicated residual DNA consists entirely of short pieces of between 0.2 and 2.2 μ in length, probably only corresponding to the "essential" sections of the chromomeres.

We may now discuss these findings in terms of the molecular aspects of chromosome organization in the polytene state. As we have seen, the amount of DNA per band per single chromatid can be calculated under the assumption that the number of strands (the degree of polyteny) remains the same along the length of a given polytene chromosome. If the further assumption is made that the relative DNA content of a given band does not, as a rule, change during development, then each chromomere (the equivalent to a band on the level of the individual chromatid) can be assigned a specific, genetically fixed DNA value. This value can be expressed in terms of numbers of nucleotides, or as length in microns of DNA double helix, with the implicit assumption that the chromomeric sections of the chromosome as well as the interchromomeric ones represent portions of a single, differentially folded, continuous strand. It has been shown that, if all these assumptions are accepted as valid, the DNA values corresponding to individual bands in either *Drosophila*, or *Chironomus*, are of the order of 20000 to 40000 nucleotide pairs on average and possibly range between about 3000 to about 100000 nucleotide pairs in individual bands. For instance, in the case of the *zeste-white* interval in the X chromosome of *Drosophila melanogaster* which, according to RUDKIN (1965) contains 0.9% of the DNA of the euchromatic portions of the X, an estimated 3×10^5 nucleotide pairs are distributed between 14 bands of various thickness to which correspond, as we have seen, 14 functional genetic units. Thus, per band, as well as per unit, there are on average 20000 nucleotide pairs of DNA, of which only a fraction seems to be essential.

Before entering upon a discussion of the genetic implications of such a situation, if it exists, it appears necessary to consider possible alternative interpretations of polytene chromosome organization. For instance, in spite of the fact that the banding pattern in various differentiated tissues remains essentially unchanged, a mechanism of differential replication of different individual chromomeres could be envisaged such that the specific DNA values of different bands would only be attained secondarily during development. One-band deficiencies, as observed in the salivary gland chromosomes, might thus sometimes be apparent rather than real, possibly indicating only a failure of the gene "magnification" process during polytenization which would not necessarily imply inactivation of this gene in relevant developmental situations. In order to achieve the variety of chromomere sizes actually observed, a maximum of only 3 to 4 extra replications at some loci would be required which on the whole would amount to no more than just one apparent extra replication of the entire genome. This would probably go unnoticed. Although no specific facts arguing against this possibility can be brought forward, and in fact a phenomenon such as the formation of DNA puffs in Sciara (CROUSE and KEYL, 1968) demonstrates that local extra replications can occur, it seems extremely unlikely that the banding pattern of giant chromosomes in general develops through this mechanism. The amazing reproducibility of the banding from cell to cell and from individual to individual would require an absurd rigidity in the control of the postulated differential replication mechanism, — absurd because this rigidity would imply a corresponding selective advantage in maintaining locus specific DNA values regardless of gene function. The proposition becomes even more absurd if it is restricted to cells and tissues which develop polytene chromosomes. Pattern rigidity to the extent actually observed in polytene chromosomes is more likely to have a structural basis, i.e., it must reflect the actual molecular subdivision of the chromosomes or, more precisely, of the DNA. The idea of differential gene "magnification" as a basis for cellular differentiation, attractive as it may appear at first sight, does not find much support in giant chromosome cytology.

In accepting that the banding pattern of giant polytene chromosomes is more or less directly representative of the linear subdivision of the genetic material itself, the chromosome can be viewed simply as a single DNA double helix or a chain of such helices which is continuous through chromomeres as well as interchromomeres, with the chromomeres representing relatively long but condensed of folded sections of the strand (Beermann, 1962). In theory, this model could be replaced by another one based on the branching of the DNA in each chromomere into many parallel identical double helices of short length. Apart from the inherent theoretical difficulties of such a model with respect to the segregation mechanism following replication and with respect to mitotic recombination, it is in conflict with the observation that single bands of polytene chromosomes can form large puffs of the Balbiani ring type (Beermann, 1957). These puffs have been found to be composed of a large number of loops each up to 10 μ in length: If in analogy to lampbrush chromosome loops the axis of these loops is taken to be DNA, the observed length can only be achieved by spinning out all or almost all of the DNA from each chromomere into one continuous DNA strand.

The conclusion that the banding pattern of polytene chromosomes and specifically the DNA values of each individual chromomere cannot in themselves be the result of a process of chromosomal differentiation during development does not necessarily make the existence of such a structural subdivision in chromosomes less interesting for the developmental biologist. The occasional occurrence of polytene, banded, chromosomes in widely divergent groups of eukaryotic organisms (Protozoa: Ammermann, 1971; plants: Nagl, 1970; Collembola: Cassagneau, 1971) as well as the universal existence of a chromomeric subdivision during the lampbrush stage in meiotic prophase demonstrate that the discontinuous distribution of genetic functions over a chain of relatively large, discrete DNA units is a fundamental principle of eukaryotic chromosome organization. The question may now be raised whether or not, and how, this organization might contribute to the superior developmental potentialities of eukaryotes as compared to those of prokaryotes, especially in terms of cellular differentiation in multicellular systems. To answer such a question, it is essential to understand precisely what genetic functions are combined in the genetic unit cytologically defined as a chromomere, or chromomere plus interchromomere. The *Drosophila* work has shown that the distribution of essential genetic functions, i.e., those whose loss is either lethal or produces significant developmental alterations, is highly non-random with respect to the chromomeric pattern: Not only is there a numerical one to one relation between such functions and polytene chromosome bands but, as shown for the *white-Notch* interval, the actual physical association of just one essential genetic function with each individual band (or band plus interband) can also be demonstrated. Data of Rudkin (1965) which point to a direct correlation between DNA content and relative mutability in different regions of the X chromosome of *Drosophila melanogaster* are not in conflict with our conclusions because over longer intervals the differences in DNA content between individual bands would cancel out each other in such a way that the relative mutabilities correspond just as well with the relative number of bands in the sections compared. Two types of models to explain the non-random localization of essential genetic functions in eukaryotic DNA along with the excessive DNA length associated with such functions have been proposed: 1) Models based on the simple reiteration of a cistron or group of cistrons in each chromomere, and 2) chromomere models combining a more complex array of genetic functions in a kind of "super-operon". Probably all the gene localization data available at present can be reconciled with either of the two types of models so that the relative merits of each specific model can only be discussed in general terms.

In its simplest form the *reiteration model* would assume that the chromomeric DNA segment consists of a number of identical and functionally equivalent copies of one cistron coordinated into one transcription unit by means of a common initiation segment. Loss or inactivation of the initiation segment would inactivate the entire unit whereas deficiencies of considerable portions of the reiterated region could probably be tolerated without visible effects on the phenotype. In order to make such a system stable in quantitative terms, the possibility of unequal crossing-over must be strictly precluded. Otherwise, the degree of reiteration in each individual chromomere would have to be assumed to be of such vital importance that variants would immediately be eliminated by selection. Lefevre (1971) has demonstrated that regions of the X chromosome containing heavy bands "exhibit more crossing over than do regions populated by equal numbers of thin, faint bands". This is not easy to reconcile with a restriction of cross-overs to specific sites, e.g., one per band, unless, of course, it is the efficiency in establishing effective pairing contacts which is correlated with local DNA content.

Qualitatively, a multiple reiterated cistron could hardly undergo coordinated evolution through mutation and should, therefore, usually represent a population of similar but not identical nucleotide sequences. To eliminate this difficulty, either the frequent recurrence of the reiteration process and the simultaneous deletion of the preceding generation of repeated segments, or a specific checking mechanism would have to be postulated that works by means of excision and repair of DNA bases to keep the repeated cistrons identical. A reiteration model of this type is Callan's "master-slave" concept (Callan and Lloyd, 1960, Callan, 1967). Callan assumes that at least once per generation, possibly during the meiotic prophase, the base sequence of all reiterated cistrons within each chromomere is checked against that of the first or master-cistron and corrected if necessary. This allows for immediate manifestation of mutations of the reiterated unit and thus eliminates one of the basic difficulties of other models based on reiteration.

The master-slave concept convincingly explains why only one essential genetic function is usually found to be associated with one chromomere. However, quite apart from the ad-hoc character of the mechanisms postulated to secure the prevalence of the master over its slaves, the model does not satisfactorily explain the extreme conservativism with which the specific degree of reiteration is maintained in each individual chromomere. This difficulty is inherent in any model based on reiteration since, as mentioned, reiteration as such would enormously facilitate the accidental increase, or decrease, of the number of reiterated units by unequal crossing-over, even if only mitotic exchanges were permitted in the slave sections. Thus one would have to invoke the existence of extraordinary selective forces to keep the number of slaves at each locus constant. To illustrate this point, it is sufficient to recall many instances of related species within the genus *Drosophila*, or the genus *Chironomus*, which must have been sexually isolated from each other for at least 10^4 to 100000 years and which do not show any significant changes in the banding pattern of their chromosomes other than those caused by rearrangements. Still, some of the *Drosophila* gene localization data reported above indicate that experimentally induced partial deletions of chromomeres can easily be tolerated. This paradox is especially hard to reconcile with the assumption of reiteration. Not only in this respect do models of the *"super-operon"* variety seem to be superior.

Chromomere models of the *"super-operon"* type cannot simply be derived by an extension of the well-known prokaryotic operon concept to the conditions in eukaryotes. It is true that, as Rudkin pointed out earlier (1965), the average DNA content of a chromomere is of the same order as that of some giant prokaryotic operons with between 10 and 20 individual cistrons plus the operator and promoter regions, but that is where the analogy ends. In order to explain, on the simple operon model, that each chromomere contains only one "functional unit", as defined by complementation, one would have to assume that either each operon contained only one "essential cistron" (loss or inactivation of all others being neither detrimental nor otherwise easy to register), or that in such a super-operon mutations in any cistron could be non-complementing with mutations in any of the other cistrons. The latter alternative is purely theoretical. The first one is hard to disprove and may constitute an attractive possibility if the distinction between essential and non-essential cistrons is based on a selective destruction of the transcription product of each operon. In other words, the essential region of an operon would be defined as that which is actually translated into protein. Scherrer (1968) postulates such a regulatory mechanism in order to explain the rapid turn-over of the greater proportion of the RNA synthesized in the nucleus. In Scherrer's view, however, the different cistrons within one transcription unit would be equivalent, in the sense that each could become the essential one in an appropriate developmental situation. This assumption again cannot be reconciled with the *Drosophila* data which show non-complementation of mutants all along putative "operons".

A clear alternative to the prokaryotic polycistronic operon would be a mono-cistronic or oligocistronic "super-operon" in which those portions of the DNA directly or indirectly involved in regulatory functions would prevail over those carrying structural information. With the aim of explaining the enormous genome size observed in the higher forms of eukaryotic organisms, even if only the so-called unique DNA sequences are considered, Britten and Davidson (1969) have developed just such a model. They assume that in eukaryotic genomes a relatively small number of individual "producer genes" (cistrons carrying structural information), of which there may be no more than exist in prokaryotic genomes, are integrated in a system of complex regulatory interactions working within the nucleus and based on the presence of a vast majority of DNA sequences exclusively involved in regulation. The expression of each producer gene would be under the control of many "receptor" genes (promoters) structurally linked to it. Transcription of the producer gene can only occur if at least one of its receptor genes is "activated" by forming a sequence-specific complex with "activator RNA". The latter RNA would be synthesized by specific integrator genes which, in turn, are thought to be under the control of "sensor genes", the latter being sensitive to developmental signals of various types. To make the model still more flexible, not only the receptor genes are assumed to be redundant but also the integrator genes so that one sensor would control the activity of many integrator genes which, in turn, could control large "batteries" of producer genes by means of redundancy in the receptor sections.

The relevant features of the Britten-Davidson model which make it attractive in the present context are two-fold. Firstly, the model allows for considerable evolutionary changes in the overall amount of regulatory sequences as opposed to the number of sequences carrying structural information. Both producer-receptor complexes and sensor-integrator complexes could also vary considerably in size within a given genome, thus explaining, if the equivalence of these complexes to chromomeres is accepted, the different DNA values of individual chromomeres. Secondly, and probably more important, the assumption of an intra-nuclear regulation mechanism based on the interaction of specific types of RNA with the DNA template could both simplify regulation as such and make it less sensitive to mutational changes since for the purported complex formation between RNA and DNA the conditions of sequence specificity might be less stringent than those required for the interaction between protein and DNA. Thus, only that portion of the chromomere carrying structural information would appear as essential in genetic complementation tests. However, one feature of the Britten-Davidson model should not be over-looked: It requires the existence of two different types of genetic functional units, as explained above, but there is no cytological or cytogenetic indication for the existence of two types of chromomeres.

A more conservative super-operon model has been proposed by GEORGIEV (1969). This model divides the operon in higher organisms into two main parts: A promoter-proximal acceptor (or non-informative zone) and a promoter-distal structural (or informative zone). Numerous "acceptor loci" in the non-informative zone are believed to interact specifically with regulatory proteins in the manner of operators, or promoters. It is believed that the informative zone not only contains cistrons for structural or enzyme proteins but also for regulatory proteins. The entire unit is believed to be transcribed in one piece with the acceptor portion being degraded before the messenger leaves the nucleus. GEORGIEV does not specify the actual mechanism underlying a multiple acceptor type of transcription control, whether it be negative or positive. In order to make such a system work, one should probably have to assume, just as in the case of the Britten-Davidson model, that transcription does not depend on the simultaneous repression or induction of all acceptor (or receptor)sequences of a given super-operon. Essentially, the Georgiev model of genomic organization in eukaryotes is an extreme variant of the Britten-Davidson system, in which the assumption that a second type of unit is devoted entirely to regulation is avoided. Each functional genetic unit is assumed to carry not only a number of acceptor sites but also a number of activator (or integrator) genes in addition to the structural cistrons. With respect to gene evolution, this would simplify matters since, with respect to the mutational tolerance of the system, the same arguments apply as in the case of the Britten-Davidson model.

With the exception of Callan's master-slave model, which is adapted specifically to the situation in lampbrush chromosomes, none of the models discussed so far have been specifically molded to fit actual cytological data such as those presented in this article. Recently, however, CRICK (1971) has proposed a model of the super-operon type that rests primarily upon polytene chromosome cytology: He takes the extreme stand of placing the structural cistrons exclusively in the interbands and leaving the DNA of the bands entirely to the task of regulation. The important point that is made in CRICK's proposal is that there may be a requirement for very long DNA base sequences in the regulatory segments of eukaryotic genes, in order to secure non-ambiguity in complex recognition functions involving DNA and proteins.

Finally, in contrast to those models of the super-operon type, in which all the DNA in the chromomere-inter-chromomere complex is thought to be engaged in activities of genetic significance, informative or regulatory, one could also conceive of a subunit organization in which genetically essential DNA carried a long tail of less essential DNA base sequences playing a mere auxiliary, secondary role in transcriptional or post-transcriptional activities. From a study of the puffing process in polytene chromosome, as well as from observations on the formation of chromomeric loops in lampbrush chromosomes, one can conclude that transcription products are not immediately released from the template but are retained, and even stored there for some time. This in itself may simply be due to the enormous length of DNA that has to be transcribed, before the transcription product is finally released into the nuclear sap. It has been proposed (BEERMANN, 1965) that the transient retention, or storage, of the transcription product in a looped or puffed site may in itself be critical for the further processing of the genetic message. The genetically relevant portion of the transcription product (p-segment) would simply be kept immobilized at its distal end as long as transcription of the nonsense portion of the template (np-segment) was still proceeding. This might facilitate, yet not always be necessary for, the complexing of the essential portion of the transcription product with a specific transport protein or, in complex situations, with two or more different proteins which might program the subsequent translational behavior of the message. This model could explain why deletion of the major portion of a band can often be tolerated by the individual, at least in the laboratory, although such a loss might confer a long-term selective disadvantage to the genotype under stringent natural conditions.

A chromomere model which shares an essential attribute with the one discussed here has recently been proposed by PAUL (1972, in press): He assumes that chromomeres originate through an initial reiteration of conventional genetic units which comprise an "address locus", in addition to the promoter, regulator, initiator, protein coding, and terminator sequences. All but the first of the reiterated units are thought to gradually degenerate during further evolution, both with respect to base sequence and to their original functions, with the exception of the "address loci" which continue to be needed as recognition sites for a postulated unwinding mechanism, plus the terminator of the last unit. A fully evolved chromomere would thus have only one "sensible" gene at its beginning, this initial segment being followed by an appendix of nonsense DNA sequences which, although being transcribed, possess only auxiliary functions.

In summarizing this discussion, the following view of the eukaryotic genome emerges: The basic structural and functional unit of the eukaryotic chromosome is the chromomere, or the chromomere-interchromomere complex. As a rule only one protein coding sequence, or cistron, seems to exist within each chromomere-inter-chromomere complex. The majority of the DNA in each chromomere and thus in the entire genome seems to be devoted to complex regulatory and control functions in addition to merely structural ones. The total number of different protein-coding cistrons may not exceed values between 5000 and 10000. The enormous variety and versatility displayed be higher organisms both in their evolution and in their individual development are thought to be based entirely on the potentialities of the intra-nuclear control system, the structural basis of which is well manifested in the banding pattern of polytene chromosomes. Developmental studies on giant polytene chromosomes signify an important second step in the "attack on the gene".

References

ALIKHANIAN, S. J.: A study of the lethal mutations in the left end of the sex chromosome in *Drosophila melanogaster*. Zool. Zhur. 16 (1937).

AMMERMANN, D.: Morphology and development of the macronuclei of the ciliates *Stylonychia mytilus* and *Euplotes aediculatus*. Chromosoma (Berl.) **33**, 209—238 (1971).

ARCOS-TERAN, L., BEERMANN, W.: Changes of DNA replication behavior associated with intragenic changes of the *white* region in *Drosophila melanogaster*. Chromosoma (Berl.) **25**, 377—391 (1968).

BAHR, G. F.: Human chromosome fibers. Considerations of DNA-protein packing and of looping patterns. Exp. Cell Res. **62**, 39—49 (1970).

BEERMANN, W.: Chromomerenkonstanz und spezifische Modifikationen der Chromosomenstruktur in der Entwicklung und Organdifferenzierung von *Chironomus tentans*. Chromosoma (Berl.) **5**, 139—198 (1952).

— Chromosomal differentiation in insects. In: (Ed.): D. Rudnick, Developmental Cytology, pp. 83—103. New York: Ronald, 1959.

— Riesenchromosomen. Protoplasmatologia. Handbuch der Protoplasmaforschung, Bd. IV d. Wien: Springer 1962.

— BAHR, G. F.: The submicroscopic structure of the Balbiani-Ring. Exp. Cell Res. **6**, 195—201 (1954).

BERENDES, H. D.: Polytene chromosome structure at the submicroscopic level: I. A map of region X, 1—4 E of *Drosophila melanogaster*. Chromosoma (Berl.) **29**, 118—130 (1970).

BRIDGES, C. B.: Salivary chromosome maps. J. Heredity **26**, 60—64 (1935).

— A revised map of the salivary gland X chromosome of *Drosophila melanogaster*. J. Heredity **29**, 11—13 (1938).

BRITTEN, R. J., DAVIDSON, E. H.: Gene regulation for higher cells: A theory. Science **165**, 349—357 (1969).

CALLAN, H. G.: The organization of genetic units in chromosomes. J. Cell Sci. **2**, 1—7 (1967).

— LLOYD, L.: Lampbrush chromosomes of crested newts *Triturus cristatus* (Laurenti). Phil. Trans. Roy. Soc. London, B **243**, 135—219 (1960).

CASSAGNEAU, P.: Les chromosomes salivaires polytènes chez *Bilobella grassii*. Chromosoma (Berl.) **35**, 57—83 (1971).

CRICK, F. H.: General Model for the Chromosomes of Higher Organisms. Nature (Lond.) **234**, 25—27 (1971).

CROUSE, H. V., KEYL, H. G.: Extra replications in the "DNA-puffs" of *Sciara coprophila*. Chromosoma (Berl.) **25**, 357—364 (1968).

DEMEREC, M.: The nature of changes in the *white-Notch* region of the X chromosome of *Drosophila melanogaster*. Proc. 7th Int. Genet. Congr. 1939. J. Genet. Suppl. **1941**, 99—105.

EDSTRÖM, J. E.: Chromosomal RNA and other nuclear fractions. In: Role of chromosomes in development, pp. 137—152. New York: Academic Press 1964.

ENGSTRÖM, A., RUCH, F.: Distribution of mass in salivary gland chromosomes. Proc. Nat. Acad. Sci. (Wash.) **37**, 459—461 (1951).

GEORGIEV, G. P.: On the structural organization of operon and the regulation of RNA synthesis in animal cells. J. theor. Biol. **25**, 473—490 (1969).

GERSH, E. S.: A new locus in the *white-Notch* region of the *Drosophila melanogaster* X chromosome. Genetics **51**, 477—480 (1965).

— Genetic effects associated with band 3C1 of the salivary gland X chromosome of *Drosophila melanogaster*. Genetics **56**, 309—319 (1967).

Green, M. M.: Putative non-reciprocal crossing-over in *Drosophila melanogaster*. Z. Vererb.-Lehre **90**, 375—384 (1959).
— Variegation of the eye color mutant *zeste* as a function of rearrangements at the *white* locus in *Drosophila melanogaster*. Festschr. Stubbe, Biol. Zbl. Suppl., 211—220 (1967).
— The genetics of a mutable gene at the *white* locus of *Drosophila melanogaster*. Genetics **56**, 467—482 (1967).
Heitz, E.: Über α- und β-Heterochromatin sowie Konstanz und Bau der Chromomeren bei *Drosophila*. Biol. Zbl. **54**, 588—699 (1934).
— Bauer, H.: Beweise für die Chromosomen-Natur der Kernschleifen in den Knäuelkernen von *Bibio hortulanus* 1. Z. Zellforsch. **17**, 68—82 (1933).
Hochman, B.: Analysis of chromosome IV in *Drosophila melanogaster*. II: Ethyl methanesulfonate induced lethals. Genetics **67**, 235—252 (1971).
Judd, B. H.: Analysis of products from regularly occurring asymmetrical exchange in *Drosophila melanogaster*. Genetics **46**, 1687—1697 (1961).
— Shen, M. W., Kaufmann, Z. C.: The anatomy and function of a segment of the X chromosome of *Drosophila melanogaster*. Genetics **71**, in press (1972).
King, R. L., Beams, H. W.: Somatic synapsis in *Chironomus*, with special reference to the individuality of the chromosomes. J. Morph. **56**, 577—586 (1934).
Konopka, R. J., Benzer, S.: Clock mutants of *Drosophila melanogaster*. Proc. nat. Acad. Sci. (Wash.) **68**, 2112—2116 (1971).
Lefevre, G.: Report on new mutants. D.I.S. **26**, 66 (1952).
— Salivary chromosome bands and the frequency of crossing over in *Drosophila melanogaster*. Genetics **67**, 497—513 (1971).
— Wilkins, M. E.: Cytogenetic studies on the *white* locus in *Drosophila melanogaster*. Genetics **53**, 175—187 (1966).
— Green, M. M.: Interactions of deficiencies in the 3C region. D.I.S. **46**, 141 (1971).
Lewis, E. B.: Genes and developmental pathways. Amer. Zool. **3**, 33—56 (1963).
Lifschytz, E., Falk, R.: Fine structure analysis of a chromosome segment in *Drosophila melanogaster*. Analysis of X-ray induced lethals. Mutation Res. **6**, 235—244 (1968).
— — Fine structure analysis of a chromosome segment in *Drosophila melanogaster*. Analysis of ethyl methanesulphonate-induced lethals. Mutation Res. **8**, 147—155 (1969).
Lindsley, D. L., Grell, E. H.: Genetic variations of *Drosophila melanogaster*. Carnegie Institution of Washington Publication No. 627 (1968).
Lowman, F. G.: Electron microscope studies of *Drosophila* salivary gland chromosomes. Chromosoma (Berl.) **8**, 30—52 (1956).
Mackensen, O.: Locating genes on salivary chromosomes. J. Heredity **26**, 163—174 (1935).
Mulder, M. P., van Duijn, P., Gloor, J. H.: The replication organization of DNA in polytene chromosomes of *Drosophila hydei*. Genetica **39**, 385—428 (1968).
Nagl, W.: 4096-ploidie und Riesenchromosomen im Suspensor von *Phaseolus collineus*. Naturwissenschaften **49**, 261—262 (1962).
Painter, T. S.: A new method for the study of chromosome rearrangements and the plotting of chromosome maps. Science **78**, 585—586 (1933).
— Salivary chromosomes and the attack on the gene. J. Heredity **25**, 465—476 (1934).
Panshin, I. B.: Cytogenetic analysis of the homology of genes in reversed linear repeats. C.R. Acad. Sci. USSR **30**, 57—60 (1941).
Prescott, D. M., Bostock, C. J., Murti, K. G., Lauth, M. R., Gamow, E.: DNA of ciliated Protozoa. 1. Electron microscopic and sedimentation analysis of macronuclear and micronuclear DNA of *Stylonychia mytilus*. Chromosoma (Berl.) **34**, 355—366 (1971).
Rasch, E. M., Barr, H. J., Rasch, R. W.: The DNA content of sperm of *Drosophila melanogaster*. Chromosoma **33**, No. 1, 1—18 (1971).
Rayle, R. E.: Mutation and chromosomal puffing in the vicinity of the *white* locus of *Drosophila melanogaster*. Dissertation Univ. of Illinois 1967.
— Green, M. M.: A contribution to the genetic fine structure of the region adjacent to *white* in *Drosophila melanogaster*. Genetica **39**, 497—507 (1968).
Ris, H., Kubai, D. F.: Chromosome structure. Annual Rev. of Genetics **4**, 263—294 (1970).

RUDKIN, G. T.: Cytochemistry in the ultra-violet. Microchem. J. Symp. Ser., **1**, 261—278 (1961).
— The relative mutabilities of DNA in regions of the X chromosome of *Drosophila melanogaster*. Genetics **52**, 665—681 (1965).
— CORLETTE, S. L., SCHULTZ, J.: The relations of the nucleic acid content in salivary gland chromosome bands. Genetics **41**, 657—658 (1956).
SCHALET, A., LEFEVRE, G., SINGER, K.: Preliminary cytogenetic observations on the proximal euchromatic region of the X chromosome of *D. melanogaster*. D.I.S. **45**, 165 (1970).
SCHERRER, K., MARCAUD, L.: Messenger RNA in Avian erythroblast at the transcriptional and translational levels and the problem of regulation in animal cells. J. Cell Physiol. **72**, Suppl. 1, 181—212 (1968).
SLIZYNSKA, H.: Salivary chromosome analysis of the white-facet region of *Drosophila melanogaster*. Genetics **23**, 291—299 (1938).
SORSA, M.: Ultra-structure of the polytene chromosome in *Drosophila melanogaster* with special reference to electron microscopic mapping of chromosome 3R. Ann. Acad. Sci. Fenn. Ser. A, IV. **151**. 1—18 (1969).
—, SORSA, V.: Electron microscopic observations on interband fibrils in *Drosophila* salivary chromosomes. Chromosoma (Berl.) **22**, 32—41 (1967).
—, SORSA, V.: Electron microscopic studies on band regions in *Drosophila* salivary chromosomes. Ann. Acad. Sci. Fenn. Ser. A IV. **127**, 1—8 (1968).
SORSA, V., SORSA, M.: Ideas on the linear organization of chromosomes revived by electron microscopic studies of stretched salivary chromosomes. Ann. Acad. Sci. Fenn. Ser. A. IV. **134**, 1—16 (1968).
SUTTON, E.: A cytogenetic study of the *yellow-scute* region of the X chromosome in *Drosophila melanogaster*. Genetics **28**, 210—217 (1943).
SWIFT, H.: Nucleic acids and cell morphology in Dipteran salivary glands. In: Allen, J. M. (Ed.): The Molecular Control of Cellular Activity, pp. 73—125. New York: McGraw-Hill 1962.
WELSHONS, W. J.: Analysis of a gene in *Drosophila*. Science **150**, 1122—1129 (1965).

Chromosomes Isolated from Unfixed Salivary Glands of Chironomus

Markus Lezzi and Michel Robert

Zoologisches Institut der ETH,
Labor für Entwicklungsbiologie, Zürich

Institut für Genetik der Universität des Saarlandes, Saarbrücken

I. Introduction

The strategy of cell differentiation is based primarily on the principle of differential gene activation. With any mechanism of differential gene activation, the final processes leading to the activity of a gene occur at the level of the chromosome. Discussion about the probable nature of these final processes, and the intermediates involved therein, will end with experiments in which such processes prove to be operational in isolated chromosomes. Using regular chromosomal material, the gene specificity of an activation process is difficult to establish. However, with polytene chromosomes of salivary glands from *Chironomus* this can be done easily and precisely. These chromosomes exhibit a very characteristic banding pattern and are so big that they can be isolated even by freehand micro-manipulation. Thus, they represent an ideal tool for studying differential gene activation at the chromosome level.

The usefulness of isolated polytene chromosomes for such studies is just being recognized. Therefore, only preliminary results on this subject are available at the present time. Findings concerning the general structure of isolated polytene chromosomes were reported as early as 1946 (d'Angelo). Since the understanding of the chemical, structural, and functional properties of isolated polytene chromosomes is regarded as a prerequisite for the understanding of gene regulation at the chromosome level, a substantial part of the following article is devoted to these general aspects.

II. Methods

A. Isolation of Single Chromosomes

The method introduced by d'Angelo (1946) is the basis of all other methods using freehand micro-manipulation as an approach to chromosome isolation. In principle, this method consists of the following steps: 1) Explant the salivary gland in isolation medium. 2) With a metal needle make a slit in the cytoplasm of a cell in order to open a passageway for the chromosomes. 3) Tear open the nuclear membrane. 4) Quickly remove the chromosomes from the nucleus.

Variations concern the type of isolation media used, the transport of chromosomes, and the set-up for incubation and observation of the chromosomes.

3*

1. Isolation Media

D'Angelos 90 mM KCl plus 60 mM NaCl solution at pH 7 is still the best medium for isolating unfixed chromosomes with ease. It has the disadvantage of dissolving the salivary gland secretion which could contaminate the chromosomal preparation. Puff material might be lost because it becomes very loose in this medium. The addition of calcium or magnesium ions to the medium makes the secretion more viscous, the cytoplasm harder and the chromosomes more rigid. If monovalent cations must be avoided, a 200 mM sucrose solution containing 10 mM $MgCl_2$ is recommended. Addition of 1 mM $CaCl_2$ appears to change favorably the consistency of the cytoplasm and the cell membrane.

2. Transport

Isolated chromosomes can be transferred to another drop by the use of a micro-pipette (D'Angelo, 1946). Dislocation of chromosomes within the same drop is accomplished very easily and quickly with a fine needle to which the chromosomes are attached. (Lezzi and Gilbert, 1970).

3. Incubation

Deposition of the chromosomes in a micro-incubation chamber proved to be very useful for experiments with repeated changes of media. The micro-incubation chamber consists of a hole in a glass (Gruzdev and Belaja, 1968a) or Parafilm (Lezzi and Gilbert, 1970) sheet sealed to a slide. When the drop of medium on top of the chamber is exchanged, the new solution mixes immediately with the remainder of the old solution left in the chamber without disturbing the chromosomes. If the incubation droplet has to be kept small and if incubation proceeds for an extended period of time, paraffin oil may be layered over the preparation in order to avoid evaporation (Lezzi, 1967a). In these cases also the isolation of the chromosomes is conducted under the oil. With the oil-covered preparation the medium cannot be exchanged, although substances can be added to the medium through the oil. High-power magnifications cannot be used.

4. Observation

The methods described below are especially designed for observations of unfixed chromosomes. Thus, they permit the investigator to follow experimentally induced changes of chromosome structure "live". Phase-contrast illumination yields very clear pictures (Figs. 4, 11, 13) if the difference in the refractive indices of chromosome and medium is not too high and if the depth of the preparation is kept small and constant by the use of a coverslip. A slight closure of the condenser diaphragm produces a phase-contrast-like picture even under suboptimal optical conditions, e.g. by immersing the objective lens directly in the incubation medium (Figs. 8, 9, 10). This type of observation is most convenient. Oblique illumination very impressively reveals the three-dimensional surface structure of chromosomes (Figs. 5, 6). A reasonable application of this type of observation is restricted to low-power objective lenses. The same restriction applies to the use of interference contrast optics which were not found to be superior to oblique illumination.

5. Cytochemical Tests

Methods for fixation of isolated chromosomes and for preparation of autoradiographs are described elsewhere (LEZZI and GILBERT, 1970). Here only the staining reaction for proteins will be considered. Since the specificity of the fast green reaction at a controlled pH is very often overestimated the conditions are given (Table 1) under which different proteins can be more or less specifically distinguished.

Table 1. Staining reactions for different proteins

Reaction No.	pH of fast green staining	Components previously extracted	Specificity
1	8	—	basic non-histone proteins
2	8	DNA	all basic proteins
3	8	DNA plus basic non-histone proteins	histones
4	2	—	all proteins
5	2	all basic proteins	acidic-neutral proteins

B. Mass Isolation

Methods for mass isolation of polytene chromosomes will have a great impact on biochemical studies of chromosomes. So far, only a rather unsatisfactory method has been published which makes use of proteinase (KARLSON and LOEFFLER, 1962). Recently ROBERT (1971) developed a more gentle method for mass isolation of polytene chromosomes. This method makes use of a detergent instead of proteinase and proceeds in two steps: 1) clean nuclei are isolated; 2) the nuclear membrane is disrupted mechanically and the chromosomes are freed.

1. Isolation of Nuclei

a) Soaking of Glands in Detergent

Digitonin is a non-ionic detergent of the steroid type. It has the useful property of being able to dissolve the cell membrane while leaving the nuclear membrane intact. For isolation of salivary gland nuclei it was found to be superior to the commonly used detergents like Triton X-100 or Tween. Soaking in detergent takes about 40 min and is done at 2° C.

b) Composition of Isolation Medium

The presence of small amounts of Ca^{2+} (1.3 mM) during the whole isolation procedure of nuclei and chromosomes is indispensable for long-lasting preservation of an intact chromosome structure. A slightly acidic pH (pH 6.3) prevents aggregation of the nuclei. Presence of sucrose should be avoided during the isolation as it makes the cytoplasm hard and difficult to remove, thus impairing the purity and yield of the nuclear preparation. However, with a Ringer-type solution, 80—100% of the nuclei of a gland can be recovered in a microscopically clean state.

2. Isolation of Chromosomes

The shearing forces which are required for breaking nuclear membranes are produced by pipetting. The speed with which the nuclei are forced through the pipette is critical, as is the inner diameter of the pipette used for this procedure: with a low speed or a wide-mouth pipette the nuclei do not break, whereas with a high speed or a narrow-mouth pipette the chromosomes become distorted. Siliconized micropipettes of appropriate diameter are also used for transferring isolated chromosomes or nuclei.

III. Results

A. Functional Properties

Are isolated chromosomes still "alive"? No general answer to this question can be given, as it depends on the particular function in question. As revealed by autoradiography, chromosomes isolated by micro-manipulation are able to synthesize RNA and perhaps to incorporate amino acids (Lezzi, 1967a). Whether they can also replicate their DNA has not yet been tested. A complete reduplication of the entire chromosome is not to be expected.

Regions of isolated chromosomes active in synthesizing RNA are the Balbiani rings and large puffs. These regions will here be called "active regions". This does not imply that all the rest of the chromosome is completely inactive. The boundary between the active and inactive states of a chromosome region is not clear. Therefore, only a band which does not show any changes in the direction of a puff will here be called "inactive".

The fact that only a few regions of isolated chromosomes appear to be active in synthesizing RNA could be due to 1. a locally restricted unmasking of RNA polymerase which would be evenly distributed along the chromosome; 2. a local accumulation of RNA polymerase at the active regions. The first possibility can be rejected because of the observation made on chromosomes partially deprived of the masking principle, the histones. These chromosomes do not show any extension of the RNA-synthesizing areas (Fig. 1). (It is notable that DNA-bound RNA polymerase is very stable in the presence of Mg^{2+}.) The above statement concerning the preservation of the activity pattern in partially dehistonized chromosomes holds true only if endogenous, i.e. bound RNA polymerase is used as a catalyst of RNA synthesis. It does not hold true if an excess of *free* RNA polymerase is provided in a

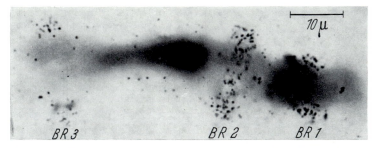

Fig. 1. Endogenous RNA-synthesizing activity pattern of isolated chromosome IV (*C. tentans*) after partial dehistonization with trypsin. Autoradiograph. Radioactive precursor: ³H—CTP (From Lezzi, 1967a)

soluble form. In contradistinction to measurements of the *endogenous RNA-synthe-sizing activity*, the template activity of a chromosome is measured with the addition of extra polymerase.

The *template activity* indicates which regions of the chromosome are accessible for free RNA polymerase molecules such that RNA synthesis can be initiated. In untreated isolated chromosomes the regions exhibiting high endogenous activity coincide with those of high template activity. However, in partially dehistonized chromosomes (see section III.B.1.a) β) 1) the template activity is greater than the endogenous activity as it extends beyond the originally active sites (LEZZI and GILBERT, 1970). Template activity can, and preferably should, be measured after the endogenous polymerase activity has been destroyed.

B. Structural Properties

1. General

a) Inactive Regions

α) **Longitudinal Integrity.** By use of a micro-manipulator D'ANGELO (1946) could stretch isolated polytene chromosomes longitudinally up to 25-fold without inducing any breaks. Release of the tension caused the chromosomes to return almost to their original length. Lateral stretching splits the chromosomes into individual fibers which carry chromomere-like nodules at the sites corresponding to the bands. D'ANGELO's findings clearly demonstrated that polytene chromosomes represent a bundle of fibers. The individual fibers are of great elasticity and strength. The elasticity is reduced at high pH and increased in the presence of calcium ions (D'ANGELO, 1946). The strength of the fibers obviously derives from their DNA rather than their protein moiety: DNases but not trypsin or chymotrypsin are able to break isolated polytene chromosomes into small pieces, the size of a few bands (LEZZI, 1965; LEZZI as cited by ALLISON, 1967).

β) **Lateral Integrity.** Theoretically, the lateral integrity depends on two types of forces: inter- and intra-chromatidial forces. The intra-chromatidial forces are primarily those which keep a chromomere in a compact and condensed state. Natur-ally, these forces also affect the *length* of the chromatidial fiber. In practice it is very difficult to distinguish between inter- and intra-chromatidial forces. Therefore, only the lateral integrity of a band as a whole is considered, although it is assumed that most of the effects described below essentially concern the intra-chromatidial forces.

1. Effect of Ions. Starting from an ion-free solution (e.g. distilled water), an increase in the ionic strength of the medium affects the lateral integrity of a chromo-some in two opposite directions: up to a critical salt concentration the diameter of the chromosome is reduced; above this concentration the diameter is increased again (GRUZDEV and BELAJA, 1968a). In the most contracted state a chromosome looks almost normal, i.e., it shows a clear banding pattern. The concentration at which the chromosomal diameter is smallest ranges for NaCl or KCl from 125 to 150 mM, for $MgCl_2$ or $CaCl_2$ from 5 to 50 mM (Fig. 2). For the sake of clarity, salt solutions of these concentrations are referred to in the following sections as "isotonic". The chromosomal expansion occurring at very low ionic strengths is fundamentally different from that occurring at very high ionic strengths: Unless divalent cations are removed (e.g. by EDTA), the first type of expansion is reversible whereas the second type is not. Returning from concentrations of 200−400 mM NaCl to a concentration of 150 mM restores a normally banded chromosome. However, the diameter of this chromosome is smaller than it was before treatment with high ionic strengths. (The same phenomenon is observed with a return from high to intermediate MgCl₂ concentrations.) Once the critical level of 600 mM NaCl or KCl is exceeded, a return to any lower NaCl or KCl concentration fails to reveal the banding of the chromosome. It is only upon a return to 10−12 mM MgCl₂ that the banding reappears (Lezzi and Gilbert, 1970). This finding points to an intrinsic difference between the effects of monovalent and divalent cations on the lateral integrity of

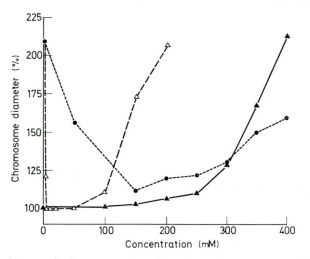

Fig. 2. Relation between ionic composition of medium and chromosome diameter. Values expressed as percentage of diameter in Mg²⁺-containing isolation medium. △ − −△: MgCl₂ solution; ●−−●: NaCl solution; ▲——▲: NaCl solution containing 12 mM MgCl₂. All solutions at pH 7.6 (From Lezzi and Gilbert, 1970)

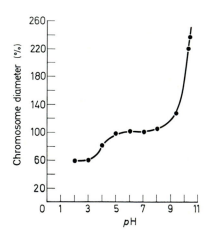

Fig. 3. Relation between pH of medium and chromosome diameter. Values expressed as percentage of diameter at pH 7.0. All solutions contain 125 mM NaCl. (From Gruzdev and Belaja, 1968a)

chromosomes. Such a difference is already suggested by the fact that untreated chromosomes are more contracted at 5—40 mM MgCl than at any of concentration NaCl or KCl (Fig. 2).

The staining reaction for histone (No. 3, Table 1) is positive with chromosomes exposed to 300—600 mM NaCl. However, a partial loss of histone cannot be detected by this method. In fact, extraction of very lysin-rich histone is assumed to have occurred in such treated chromosomes. This assumption is based on the results of parallel experiments on isolated chromatin (for references, see Lezzi, 1970). Chromosomes incubated in high ionic strength media are template-active over their entire length (compare with section III.A).

2. Effect of pH. Varying the pH of a 125 mM NaCl solution affects the chromosomal diameter, as shown in Fig. 3 (see also Fig. 4). Gruzdev and Belaja (1968a)

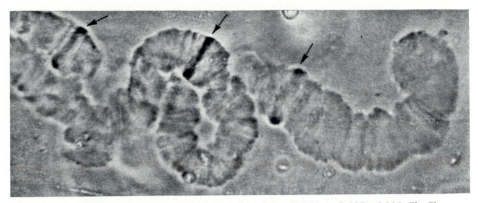

Fig. 4. General decondensation of bands induced by pH 11 and 125 mM NaCl. Chromosome of *C. thummi* (in the original paper erroneously classfied as *C. dorsalis*). — Note that the material in puff regions (arrows) stays compact (From Gruzdev and Belaja, 1968a)

stated that the swelling at an alkaline pH is irreversible; at pH > 11 the chromosomes are completely dissolved. These authors also reported that returning from a pH below 3 to pH 7 causes the heavily contracted chromosome to regain its normal appearance. This recovery process takes about 10 min, gradually progressing from the outside to the inside of the chromosome. Lezzi (unpubl. res.) found that treatment with 0.2 N H_2SO_4 alters an isolated chromosome irreversibly in that, after returning to a neutral medium, the chromosome swells vastly beyond its original diameter. The presence of 12 mM $MgCl_2$ at pH 7.6 cannot reduce this swelling. The staining reaction for total proteins (Nr. 4, Table 1) is only faint in the inactive regions of chromosomes treated in this way.

3. Effect of Enzymes. RNase and DNase have no effect on the lateral integrity of chromosomes. Trypsin or chymotrypsin cause the chromosome to expand laterally. In the presence of magnesium ions, the lateral expansion does not usually exceed twice the original diameter of the chromosome (Fig. 5). Inactive regions of such treated chromosomes stain just slightly with the reaction for total protein (No. 4, Table 1). In the absence of divalent cations, the swelling caused by trypsin or chymotrypsin continues until the chromosome is completely invisible. A return to a

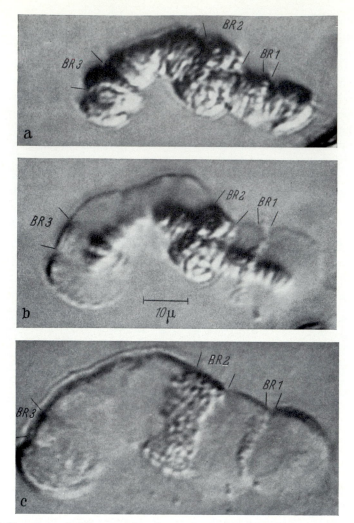

Fig. 5. Stability of rough BR material in the presence of Mg^{2+}. Effect of trypsin on isolated chromosome IV of *C. tentans*. a before; b 2 min; c $2^1/_2$ min. after trypsin addition. Note that with Balbiani ring 3 (BR3) the rough material is removed at this Mg^{2+} concentration (8 mM) (From Lezzi, 1967b)

medium containing magnesium or calcium does not cause the chromosome to reappear. Only a reduction of the pH to 2 brings back a shrunken remainder of the chromosome or just a single long thread (Lezzi, 1965; 1967b).

    4. Effect of Lipids or Organic Solvents. Benzene, xylene, chloroform or amyl alcohol do not visibly alter any of the structural features of isolated chromosomes. The same holds true for isotonic salt solutions containing saturating amounts of ecdysone or a suspension of cortisol or juvenile hormone. Dioxan, ether, ethanol, or acetone cause reversible shrinkage of the chromosome (Gruzdev and Belaja, 1968b; Lezzi, unpubl. res.).

5. Effect of Detergents. Increasing concentrations of ionic detergents (such as sodium dodecyl sulfate or sodium deoxycholate, but also cortisol-21-phosphate) produce effects similar to increasing concentrations of inorganic ions: above a critical level they cause chromosomal expansion. For deoxycholate, this critical concentration level is around 2.4 mM; for cortisol-21-phosphate, it is above 1 mM. The non-ionic detergents Triton-X and digitonine apparently do not affect chromosomal structure (GRUZDEV and BELAJA, 1968b; LEZZI, 1967b, and unpubl. res.).

6. Effect of Substances Breaking Disulfide Bonds. GRUZDEV and BELAJA (1968b) concluded from their experiments with sodium sulfite and β-mercaptoethanol that disulfide bridges are involved in the lateral integrity of chromosomes.

7. Effect of Other Agents. The presence of 10 M urea makes an isolated chromosome completely invisible. After removal of urea, i.e. in a 12 mM $MgCl_2$ solution, a banded chromosome reappears. Rarely, chromosomes so treated are shredded into several longitudinal fibres (LEZZI, unpubl. res.). A similar shredding occasionally is observed when chromosomes come into contact with the interphase between paraffin oil and the aqueous incubation medium or when they are sonified (ROBERT, unpubl. observation). Such chromosomes still contain histone (LEZZI, 1967b).

b) Active Regions

α) Effect of Ions. In an isotonic NaCl—KCl solution which is devoid of divalent cations, Balbiani rings look like cotton balls (Fig. 6). In the presence of $8-12$ mM magnesium ions these loose structures collapse into compact agglomerations of granulated material having a rough surface structure (Fig. 5). This material is here called "rough BR material". As revealed by staining reaction No. 1 (Table 1) there exists in Balbiani rings and puffs a basic non-histone protein called "basic puff protein" (Fig. 7a). This protein is no longer detectable in chromosomes treated with NaCl, KCl or $MgCl_2$ solutions of high ionic strength, although the rough BR material is usually not removed by such treatments (LEZZI, 1967b, and unpubl. res.).

β) Effect of pH. The effect of low pH, both on Balbiani ring structure and on the content of basic puff protein, is comparable to that of high ionic strength. In an alkaline environment puff regions keep at least some of their chracteristic material (Fig. 4; GRUZDEV and BELAJA, 1968a; LEZZI, unpubl. res.).

γ) Effect of Enzymes. In magnesium-free isotonic salt solutions, trypsin or chymotrypsin dissolve Balbiani rings completely. In the presence of Mg^{2+} ($5-50$ mM) these proteinases usually do not alter the rough material of Balbiani rings (Fig. 5c) nor do they extract the basic puff protein. RNAse slightly diminishes the volume of Balbiani rings. A subsequent chymotrypsin treatment will then remove the rough BR material, and the basic puff protein is no longer detectable (Fig. 7b). After removal of the rough material by this double treatment, there is still material left in the Balbiani ring regions. This material, however, has a smooth surface (LEZZI, 1967b), called here "smooth BR material".

δ) Effect of Other Agents. Among the substances mentioned in sections III.B.1.a) β) 4) and 5) only sodium dodecyl sulfate affects Balbiani ring structure, in that it causes the Balbiani ring material occasionally to "melt" on the glass surface.

The agents mentioned in section III.B.1.a) β) 7) are the only ones found which are capable of stripping off both the rough and the smooth BR material without necessarily destroying the rest of the chromosome (LEZZI, 1967b, and unpubl. res.).

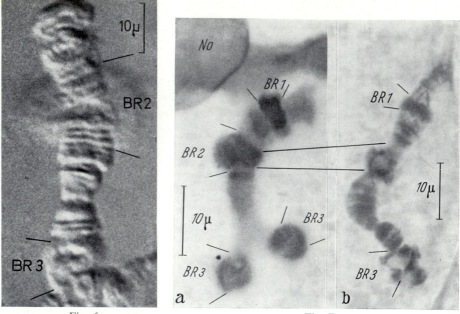

Fig. 6   Fig. 7

Fig. 6. Balbiani ring structure (BR2) in an isotonic NaCl-KCl solution free of divalent cations. Chromosome IV of *C. tentans*. (From Lezzi, 1967b)

Fig. 7. Basic non-histone protein (staining reaction No. 1, Table 1) in Balbiani rings of *C. tentans*. a without RNase pretreatment; b with RNase pretreatment, note the negative reaction in Balbiani rings 2 and 3. No: nucleolus. The chromosome in a is asynapsed in the BR3 region. (From Lezzi, 1967b)

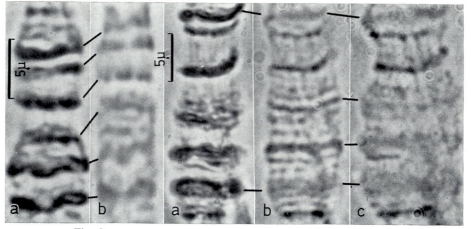

Fig. 8   Fig. 9

Fig. 8. General decondensation of bands by NaCl in the absence of Mg^{2+}. NaCl concentration of media: a 0 mM (isolation medium); b 200 mM. Stretched chromosome. (From Lezzi and Gilbert, 1970)

Fig. 9. Differential decondensation of bands by NaCl in the presence of Mg^{2+}. NaCl concentration of media: a 0 mM; b 300 mM; c 400 mM. Stretched chromosome. (From Lezzi and Gilbert, 1970)

2. Specific and Differential Properties of Chromosome Regions

The general properties described above cannot account for differences in the reponses of genes to different activating stimuli. For any mechanism of differential gene activation, the existence of specific and differential properties of chromosome regions has to be assumed. As exemplified in the following sections, differences in the properties of inactive as well as active regions do indeed exist. Some of these intrinsic differences may be utilized by the cell for differential gene activation.

a) Inactive Regions

α) Effect of Ion Combinations. The lateral expansion produced by a gradual increase from 150 mM to 300 mM NaCl or KCl is accompanied by a decondensation of the chromosomal bands. In the absence of divalent cations this decondensation progresses with all the bands of a chromosome steadily and uniformly (Fig. 8). In the presence of 12 mM $MgCl_2$, however, the decondensation is no longer homogeneous and general in that at a low Na^+ or K^+ concentration (150 mM) it affects only a small number of selected bands. The number of specifically decondensing bands increases as the NaCl or KCl titer of the Mg^{2+}-containing incubation solution approaches 300 mM (Fig. 9). The fact that under these conditions the decondensation and expansion of bands is sporadic explains why the increase in diameter of the total chromosome is less in the presence of Mg^{2+} than in its absence (Fig. 2).

Of special interest is the observation that different sets of bands decondense, whether an ion combination contains 150 mM Na^+ or 150 mM K^+. Among the bands specifically affected by a Na^+-rich ion combination are two bands in regions I-19-A; among the bands affected by a K^+-rich ion combination is a band in I-18-C of *C. tentans* (LEZZI and GILBERT, 1970).

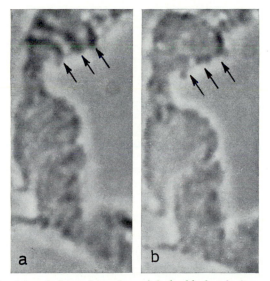

Fig. 10. Specifically delayed decondensation of 3 double bands (arrows) in chromosome region Ia4 of *C. thummi* by Mg^{2+}-rich ion combinations. Ionic composition of media: a 60 mM $MgCl_2$ plus 100 mM NaCl; b 70 mM $MgCl_2$ plus 100 mM NaCl. Note that in a decondensation has started at all the bands except the ones indicated. (Original)

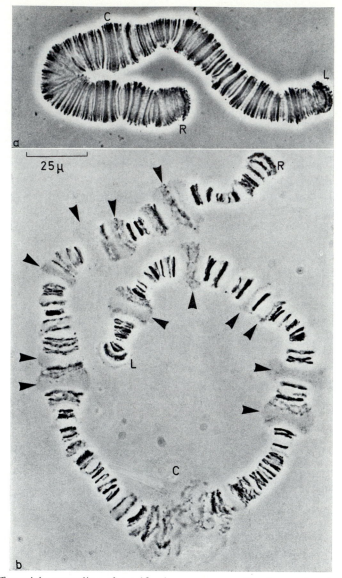

Fig. 11. Differential unravelling of specific chromosome regions by a salt solution of high ionic strength and low pH. a Control: 87 mM NaCl, 3.2 mM KCl, 1.3 mM CaCl$_2$, 1 mM MgCl$_2$, 10 mM Tris-HCl pH 7.3. b Experiment: 650 mM NaCl, 2 mM CaCl$_2$, MacIlvain buffer (4x diluted) pH 4.3. Arrows: unravelled regions. *C* centromere. *R* right arm; *L* left arm of chromosome II of *C. thummi* (From Robert, 1971)

In recent experiments (Lezzi, unpubl.res.) ion combinations were tested in which the *Mg*$^{2+}$ *titer* was varied but the NaCl concentration was kept constant at 100 mM. At 60 mM MgCl$_2$ plus 100 mM NaCl, under which condition most of the bands decondense, three chracteristic double bands in region Ia4 of *C. thummi* stay condensed. These bands decondense only at magnesium concentrations above 70 mM

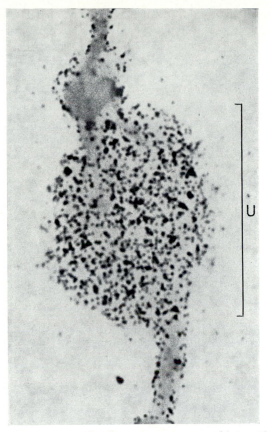

Fig. 12. Template activity pattern of a chromosome pretreated by a solution of high ionic strength (650 mM NaCl) and low pH (4.3). Autoradiograph of an unidentified segment of a chromosome of *C. thummi*. Assay medium contains extra RNA polymerase plus ATP, CTP, GTP, ³H-UTP. NaCl-concentration: 150 mM; pH 7.5. — Note that the silver are concentrated above the swollen region (U) (From ROBERT, 1971)

(Fig. 10). Subsequent studies revealed that the specificity of this effect depends only on the Mg^{2+} component of the combination since $MgCl_2$ (70—80 mM) alone produces the same effect as its combinations with NaCl. It should be re-emphasized here that, in contrast to $MgCl_2$, NaCl alone is unable to bring about even this low specificity in chromosomal band decondensation. As is evident from the findings reported next, this statement is valid only for NaCl solutions of a neutral pH.

β) **Effect of Combinations of High Ionic Strength and Low pH** (ROBERT, 1971). The results described in the following are all obtained from experiments with chromosomes isolated from the salivary glands of *C. thummi* according to the procedure of ROBERT (1971, see section II.B).

1. Local Unravelling Phenomenon in General. Whether or not divalent cations are present, 500—600 mM NaCl or KCl solutions of pH 4.3 cause a few bands of isolated chromosomes to decondense and expand to a much greater extent than do physiological salt solutions of neutral pH (see previous section). Such superpuff-like structures are shown in Fig. 11; they will be called "unravelled" regions. With increasing ionic strength the number of unravelled regions increases, comprising both potentially active bands (puff sites) as well as inactive regions such as centromeres (Fig. 11). At 800 mM NaCl or KCl all the bands of the chromosomes are unravelled. ($MgCl_2$, like NaCl or KCl, also induces unravelling although at lower concentrations, i.e. between 130 and 150 mM.) Unravelling produced by high ionic strength and low pH can survive a return to 150 mM NaCl or KCl solutions of pH 7.5.

In this physiological medium it is possible to determine the template activity pattern of chromosomes previously exposed to high ionic strength and low pH. As is evident from Fig. 12 unravelled regions incorporate more ^3H-UTP than condensed ones, provided extra RNA polymerase and the four nucleoside triphosphates are administered. If, however, the return to a physiological medium is effected in the presence of an excess of *histone* (unfractionated histone from calf thymus), the unravelling does regress to a more or less normally banded state, being no longer template-active. It is interesting to note that different histone fractions are not all equally effective in repressing unravelling and template activity.

Although the local structural changes observed with high ionic strength at low pH and those observed with low ionic strength at neutral pH differ from each other in their extent, their stability and especially their Mg^{2+} dependence, they nevertheless

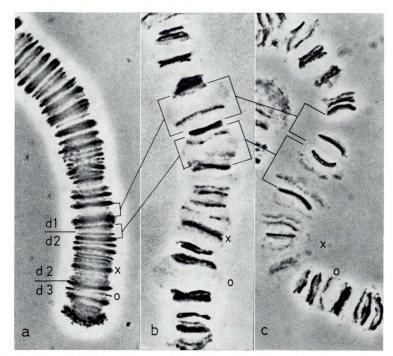

Fig. 13. Specific and differential unravelling induced by K^+ or Na^+, respectively, at high ionic strength and low pH. a Control (see legend to Fig. 11a for composition of medium). b 650 mM KCl, 12 mM $MgCl_2$, 2 mM $CaCl_2$, MacIlvain buffer (4xdiluted) pH 4.3. c same as b with 650 mM NaCl in place of KCl. Left arm of chromosome III of *C. thummi*. — Note that region IIId1 and band "b" become unravelled by K^+ b rather than by Na^+ c whereas, with region IIId1.2 and band "x", the situation is reversed. (From ROBERT, 1971)

are mutually related. As a matter of fact, these two types of structural changes can be converted in to each other by gradually increasing the pH while decreasing the ionic strength, or *vice versa*. At pH 6 and 350 mM NaCl, for example, the decondensed bands already look like small puffs.

2. Differential Ionic Sensitivity of Local Unravelling. The relationship between the unravelling phenomenon and the slight band decondensation also becomes evident from the specific ionic sensitivity of certain chromosomal regions. This may be demonstrated by the following example: In section III.B.2a) α) we mentioned that at pH 7.6 Na^+ affects region I-19-A, whereas K^+ affects region I-19-C of *C. tentans*, and that the specificity of these effects depends on Mg^{2+}. The homologous regions in *C. thummi* are IIId1.2 and IIId1, respectively. It is now very intriguing to note that at pH 4.3 it is region IIId1.2 which is primarily affected by Na^+ (650 mM), whereas region IIId1 is affected by K^+ (650 mM; see Fig. 13). Here too, the specificity depends on Mg^{2+} (6—12 mM).

Five other regions with differential ionic sensitivity have been found:

1) Region IIId2 (double band marked by "*x*" in Fig. 13) behaves like region IIId1.2, i.e. it is Na^+-sensitive.

2) Region IIId3 (band marked by "o" in Fig. 13), on the other hand, behaves like region IIId1, i.e. it is K^+-sensitive.

3) Region If4: The reacting bands in If4 are probably the site of puff formation which, in an earlier paper, was designated as "Mg^{2+}-inducible" (LEZZI, 1967c). As will be shown, this designation would be better changed to "divalent cation-inducible" since unravelling can be induced in this region, not only by Mg^{2+} (6—12 mM), but also by Ca^{2+} (2—6 mM). Both divalent cations require the presence of 650 mM NaCl in order to do this.

4) Region Igl, on the contrary, could be called a "divalent cation-repressible" region as it becomes unravelled only in the absence of Mg^{2+} and Ca^{2+}, i.e. with 650 mM NaCl plus EDTA. This region probably does not correspond to the known puff-forming site in Ig1 (KROEGER, 1963; LEZZI, 1967c).

Table 2. Differential sensitivity of chromosome regions of *C. thummi* to various combinations of Na^+, K^+, Mg^{2+}, and Ca^{2+} at different pH's

Region	Na^+	K^+	Mg^{2+}		Ca^{2+}	pH
IIId1.2	+	—	+		±	4.3 or 7.6
IIId1	—	+	+		±	4.3 or 7.6
IIId2	+	—	?		?	4.3
IIId3	—	+	?		?	4.3
If4	+	?	+	and/or	+	4.3
Ig1	+	?	—	and	—	4.3
IIe	+	?	+		—	4.3
Ia4	±	?	—		±	7.6

+ presence of respective ion required for decondensation,
— absence of respective ion required for decondensation,
± presence of absence of respective ion optional,
? effect of respective ion not yet studied.

5) Region IIe contains an unidentified band which could be classified as "Ca^{2+}-repressible", since Ca^{2+} (2—6 mM) rather than Mg^{2+} suppresses the unravelling action of 650 mM NaCl with this band.

Summarizing the above findings, one could state that in *C. thummi* we already know 8 chromosome regions which respond differentially to various combinations of Na^+, K^+, Mg^{2+}, and Ca^{2+} at different pH. These combinations are listed in Table 2. It is assumed that there exist even more regions with still other ionic sensitivities.

b) Active Regions

The combination of Na^+ or K^+ plus Mg^{2+} specifically affecting inactive bands (see section III.B.2.a) α) does not alter the appearance of active regions. The effects described below, although ion-dependent as well, clearly differ from these pure ionic effects in that they involve the action of proteinases.

In section III.B.1.a) β) 3) it was mentioned that the presence of magnesium ions prevents the rough BR material from being destroyed by trypsin of chymotrypsin. Mg^{2+} appears to stabilize this material. As is evident from Fig. 5c and the results shown in Fig. 14, the stabilizing effect of Mg^{2+} is not the same for all the Balbiani rings of *C. tentans*: BR3 requires a 10 times higher $MgCl_2$ concentration for the same degree of stability of the rough meterial than do BR1 or BR2. (For definition of stability, see legend to Fig. 14.)

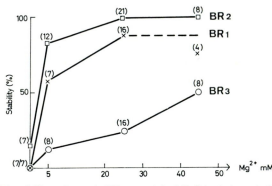

Fig. 14. Differential stability of rough BR material of Balbiani rings 1, 2 and 3 of *C. tentans*. Stability: percentage of all examined Balbiani rings whose rough BR material is not destroyed by proteinases at the Mg^{2+} concentration indicated. In parentheses: number of observations (From Lezzi, 1967b)

IV. Conclusions

A. Construction of an Inactive Region

The most important components keeping inactive bands, i.e. their chromomeres, in the compact state are the histones. Therefore only histones are shown in the model of an inactive chromomere illustrated in Fig. 15a.

Histones, particularly the very lysine-rich histone, may cross-link the coiled-up chromatidial fiber in the chromomere (see Lezzi, 1970, for references). Digestion of histones by proteinases or rupture of the ionic DNA-histone bonds by high ionic strength, or by very low or very high pH, release the chromatidial fiber from the cross-linked complex.

Thus, there exist factors which influence the DNA-histone complex and hence the stability of a chromomere. These factors are primarily pH, and the monovalent and divalent cation concentrations of the chromosomal milieu.

The *pH* of the chromosomal milieu determines the overall charge and thereby the degree of solvation and electrostatic repulsion of DNA-histone complexes. At the isoelectric point (i.e. between pH 3.3 and 3.6, DE ROBERTIS, 1960) where all the charges are neutralized, solvation and electrostatic repulsion are minimal. Therefore, at pH 3—4 nucleohistone complexes are most stable, whereas at neutral pH's (where the negative charges are in excess) the stabilization of such complexes requires the presence of extra cations.

Na[+] *and K*[+], in the concentration range and at the pH and Mg^{2+} titer present in the cell nucleus, first of all stabilize chromomeres by neutralizing the charge of their nucleohistone complex. However, under these same physiological conditions, they already start to destabilize certain chromomeres by dissociating their DNA-histone bonds.

Mg^{2+} and Ca^{2+}, at physiological concentrations (LANGENDORF et al., 1961), also contribute to charge neutralization of the nucleohistone. In addition, they appear to participate in the cross-linked DNA-histone complex itself in a manner which is not yet understood (see KABAT, 1967, for discussion of this role of Mg^{2+}). Thus, the function of Mg^{2+} and Ca^{2+} in chromomere stabilization is essentially a conservative one.

B. Construction of an Active Region

Starting from the outside of a Balbiani ring, one first finds the so-called rough material. This material can be removed from Balbiani rings by the combined action of RNase and proteinases. It contains, among others, basic protein. This basic puff protein is extracted by acid or salt[1] but is not destroyed by trypsin in the presence of Mg^{2+}. The rough BR material is believed to consist of the submicroscopic (300 A) granules found in Balbiani rings and puffs (BEERMANN and BAHR, 1954; STEVENS and SWIFT, 1966). The question of whether these granules represent clusters of informofers (GEORGIEV and SAMARINA, 1969), ribosomes (LEZZI, 1967 b) or subunits thereof attached to the growing RNA molecule, is still open to discussion. After removal of the rough BR material, the smooth BR material is found. This presumably consists of acidic-neutral protein (BERENDES, 1969). It can only be removed from the Balbiani rings by the denaturing agents sodium-dodecyl sulfate and urea, or by contact with the oil-water interphase. It is assumed that this smooth material represents, at least in part, RNA polymerase engaged in the transcription of DNA into RNA.

A model of an active chromosome region is shown in Fig. 15 d.

C. Steps in the Activation Process

The probable sequence of events leading from the inactive to the active state of a chromomere is given in Fig. 15. The point of this sequence is that the ions (Fig. 15 b) come into play prior to the RNA polymerase or any other acidic proteins (Fig. 15 c).

---

1 Earlier results contradicting this statement (LEZZI, 1967 b) subsequently could not be substantiated (LEZZI, unpubl. res.).

Ions are required for dissociation of the ionic bonds between DNA and histone. The dissociation of these bonds is a prerequisite for the binding of RNA polymerase to DNA and the initiation of RNA synthesis (see Lezzi, 1970). The possibility should not be excluded that ions also play an important role in the steps following dissociation of the DNA-histone bonds, e.g. in the interaction between growing RNA and ribosomes (Lezzi, 1967c) or informofers.

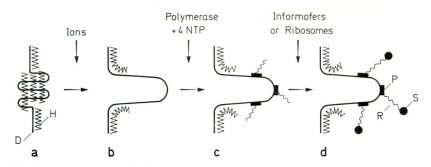

Fig. 15. Sequence of events leading from an inactive a to an active d state of a chromomere. D: DNA; H: histone; R: RNA; S: informofer or ribosome. 4 NTP: 4 ribonucleoside triphosphates. (Original)

     The fundamental problem of the differential ionic sensitivity of chromosome regions has not yet been solved. In looking for a solution of this problem the following points have to be kept in mind: 1) the ionic sensitivity is differential in two respects a) for different ionic strengths, b) for different ionic species; 2) the specificity of Na^+ or K^+ depends on the presence of divalent cations; 3) the ionic specificity does not depend on the concomitant functioning of enzymes or the addition of any other organic compound. A reasonable working hypothesis is that the chromosomes are preprogrammed in such a way that their genes react differently and specifically to the varying ionic conditions of the nucleus.

D. Biological Significance of the Differential Ionic Effects on Chromosome Regions

1. Ionic Milieu in the Cell Nucleus and Its Variation

     The exact ionic composition of the nuclear sap of *Chironomus* salivary glands is not yet known. Preliminary measurements by Kroeger (personal communication) indicate the Na^+ plus K^+ concentration of freshly isolated nuclear contents to be around 170 mM (± 50 mM). This result is in keeping with other findings on the nuclear ionic milieu (for references see Lezzi, 1970). We therefore consider a monovalent cation concentration in the range 120–220 mM to be a physiological environment for chromosomes. This is the concentration range within which the differential effects of Na^+ and K^+ on isolated chromosomes were observed (see section III.B.2.a).

     However, in contrast ot these *in vitro* conditions, the physiological milieu in the cell nucleus is very unlikely ever to be devoid of either all Na^+ or all K^+. Both these ions are probably always present together, although in different ratios. Measurements of the electropotential difference between the intranuclear and extracellular fluid of salivary glands indicate that variations in the nuclear Na^+/K^+ ratio do in fact occur during normal development, as well as after experimental hormone administration; juvenile hormone would shift this ratio up and ecdysone would shift it down (BAUMANN, 1968; KROEGER, 1966). The question now is whether such rather small variations in the Na^+/K^+ ratio can produce the same differential effects on specific band decondensation as salt solutions of extreme Na^+/K^+ ratios. The model developed in the Appendix suggests that this may well be the case.

2. Connection between in vitro Effects of Ions and in vivo Effects of Hormones

In section IV.C. it was anticipated that a band which decondenses by the action of a specific ionic combination then becomes available for transcription. This could indeed be demonstrated for ion combinations of low pH. With physiological ionic conditions, that is, at neutral pH, the resolution power of autoradiography allowed us to draw such a conclusion only for groups of several bands rather than for single bands. However, the following findings concerning the effect of Na^+ and K^+ on isolated nuclei (LEZZI, 1967c) indicate that the decondensation of a specific band is indeed related to the actual puffing activity of the corresponding site. If isolated salivary gland nuclei of *C. tentans* are incubated in a Na^+-rich medium, puffs are formed at the same two bands (region I-19-A) which become decondensed by Na^+ in isolated chromosomes. Incubation of nuclei in a K^+-rich medium causes the formation of a puff in the K^+-sensitive band in I-18-C of *C. tentans*. It is now very intriguing to note that juvenile hormone, when injected into prepupae, also brings about the formation of the two puffs in I-19-A, whereas injection of ecdysone into larvae induces the formation of the puff in I-18-C (CLEVER, 1961; LEZZI and GILBERT, 1969). As juvenile hormone ad ecdysone appear to increase or decrease, respectively, the nuclear Na^+/K^+ ratio, the following sequences of events suggest themselves quite naturally:

juvenile hormone → nuclear Na^+/K^+ up → activity in chromosome region I-19-A,

ecdysone → nuclear Na^+/K^+ down → activity in chromosome region I-18-C.

V. Appendix: A Model which Formulates the Specificity of Ion Combinations in Differential Gene Activitation

(Originally proposed by ROBERT)

This model is based exclusively on observations with isolated chromosomes which were treated with salt solutions containing as monovalent cations either sodium ions alone potassium ions alone. The results most relevant to this model are:

1) There exist *Na$^+$-sensitive chromosome regions* (e.g. I-19-A in *C. tentans*, IIId1.2 or IIId2 in *C. thummi*) which in the presence of Mg^{2+} and Ca^{2+} are activated primarily by sodium ions, and there exist *K$^+$-sensitive regions* (e.g. I-18-C in *C. tentans*, IIId1 or IIId3 in *C. thummi*) which in the presence of Mg^{2+} and Ca^{2+} are activated primarily by potassium ions.

2) Below a critical concentration (i.e. 150 mM at pH 7.6 or 550 mM at pH 4.3) neither type of chromosome region is activated by Na^+ or K^+.

3) Above another critical concentration (i.e. 225 mM at pH 7.6 or 700 mM at pH 4.3) both regions are activated by either ion.

Now, the following model tries to predict how the Na^+-sensitive and the K^+-sensitive regions would behave if chromosomes were treated with salt solutions containing *both* sodium and potassium ions in *varying combinations*.

First we ask: at any given concentration of Na^+ ($[Na^+]$) or K^+ ($[K^+]$), which is the minimal total monovalent cation concentration ($[Na^+ + K^+]_{min}$) required to activate a Na^+-sensitive or a K^+-sensitive region, respectively? This question will be discussed here for a typical Na^+-sensitive region (e.g. I-19-A); with a K^+-sensitive region (e.g. I-18-C) the situation would be essentially the same, with K^+ taking the role of Na^+, and *vice versa*. Combination A (Fig. 16) represents an extreme case in which $[K^+]$ is zero; therefore its $[Na^+ + K^+]_{min}$ equals $[Na^+]$, which has been determined to be 150 mM. Combination E represents another extreme case in which $[Na^+]$ is zero; therefore its $[Na^+ + K^+]_{min}$ equals $[K^+]$, which has been determined to be 225 mM. B and D are randomly selected combinations (for discussion of C, see below). Both have a $[Na^+]$ which, alone, is subminimal for activation of the Na^+-sensitive region. The Na^+-deficit of these combinations has therefore to be compensated by the addition of K^+. The compensatory amounts of K^+ (e.g. stretch "b" in B) need to be bigger than the deficient amounts of Na^+ (e.g. stretch "a" in B) because we have learned from combinations E and A — in which these ions are not combined — that K^+ is less effective than Na^+ in activating the Na^+-sensitive region. We expect that the relative effectiveness of K^+ and Na^+ is the same with any combination of these ions, so that $[Na^+ + K^+]_{min}$ comes always to lie on a straight line (line x). The chracteristics of combination B are: $[Na^+] = 120$ mM; $[Na^+ + K^+]_{min} = 165$ mM; those of combination D are:

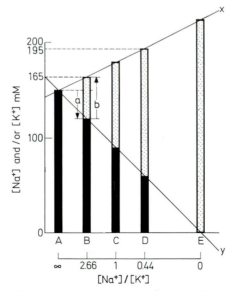

Fig. 16. Activation of a Na^+-sensitive region by solutions of a varying $[Na^+]/[K^+]$ and a *minimal* $[Na^+ + K^+]$. Five examples of possible combinations are given: A, B, C, D, E. Line x confines the minimal $[Na^+ + K^+]$ at any $[Na^+]$ given by line y. ■ = $[Na^+]$, ▦ = $[K^+]$ of the particular combination. For further explanations, see text

$[Na^+] = 60$ mM; $[Na^+ + K^+]_{min} = 195$ mM. Thus, combination D, in contrast to B, contains more K^+ than Na^+. It is obvious that combination D activates not only the Na^+-sensitive but also the K^+-sensitive region.

Therefore, we ask next: which combinations activate exclusively the Na^+-sensitive region; and which ones activate exclusively the K^+-sensitive region? If

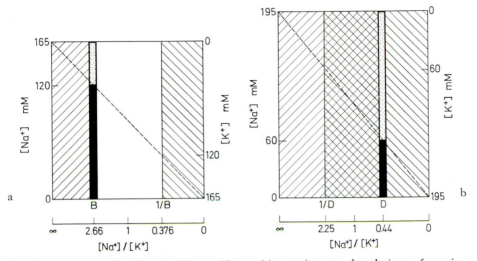

Fig. 17. Response of a Na^+-sensitive or K^+-sensitive region to salt solutions of varying $[Na^+]/[K^+]$ and *constant* $[Na^+ + K+]$. Combinations B (Fig. 17a) and D (Fig. 17b) are from Fig. 16. Combinations 1/B and 1/D have the reciprocal $[Na^+]/[K^+]$ to the ones of B and D, respectively. ■ = $[Na^+]$; ▨ = $[K+]$; ▨ = exclusive activity in the Na^+-sensitive region; ▨ = exclusive activity in the K^+-sensitive region; ☐ = no activity in either region; ▨ = activity in both regions. For further explanations

$[Na^+ + K^+]$ is kept *constant* (disregarding whether it becomes subminimal or redundant), generally two types of situation will arise, which can be exemplified by the situations shown in Fig. 17a and 17b. In Fig. 17a $[Na^+ + K^+]$ equals 165 mM which is the $[Na^+ + K^+]_{min}$ of combination B (Fig. 16). With this situation, the area of activity in the Na^+-sensitive region is confined to lower ratios by the $[Na^+]/[K^+]$ exhibited by combination B. Below that ratio there is no activity in the Na^+-sensitive region because $[Na^+ + K^+]$ of these combinations becomes subminimal with regard to their $[Na^+]$ (cf. Fig. 16). However, by further decreasing the $[Na^+]/[K^+]$ we enter an area in which the K^+-*sensitive* region becomes active. This area starts with that combination whose $[Na^+]/[K^+]$ is the reciprocal to that of B (i.e. at 1/B). In the situation shown by Fig. 17b $[Na^+] + [K^+]$ equals 195 mM, which is the $[Na^+ + K^+]_{min}$ of combination D in Fig. 16. For the discussion of this situation, we refer to the discussion of Fig. 17a (see above). The interesting point of the situation shown in Fig. 17b is that the area of no activity is absent but that there is instead an area in which *both* regions are active simultaneously.

One can easily imagine a third situation in which there is neither an area of overlapping nor an area of absent activity in the Na^+-sensitive and K^+-sensitive regions. With such an ideal situation, where the areas of exclusive activity in the

Na$^+$- or K$^+$-sensitive regions are directly adjacent to each other, it would take an infinitesimal change in the [Na$^+$]/[K$^+$] to switch the Na$^+$-sensitive region off and the K$^+$-sensitive region on, or *vice versa*. This situation would be characterized by a [Na$^+$] which equals [K$^+$] and by a [Na$^+$ + K$^+$] which is the [Na$^+$ + K$^+$]$_{min}$ of combination C (see Fig. 16). Combination C contains 90 mM each of NaCl and KCl.

The most illuminating outcomes of the model developed above are:

1) Selective activation of Na$^+$-sensitive or K$^+$-sensitive chromosome regions can be achieved by salt solutions containing both Na$^+$ and K$^+$ together.

2) The total monovalent cation content of such solutions is within the physiological range, i.e. between 120 and 220 mM (see section IV.D.1).

3) It requires a minimal change in the [Na$^+$]/[K$^+$] ratio to switch from exclusive activity in Na$^+$-sensitive regions to exclusive activity in K$^+$-sensitive regions.

Acknowledgement

We are very grateful to Miss B. Savage for help with the preparation of the manuscript and to Drs. Gruzdev and Belaja for providing us with an English translation of their article (1968b).

References

Allison, A.: Lysosome and disease. Sci. Amer. **217** (5), 62—72 (1967).

Baumann, G.: Zur Wirkung des Juvenilhormons: Elektrophysiologische Messungen an der Zellmembran der Speicheldrüse von *Galleria mellonella*. J. Insect. Physiol. **14**, 1459—1476 (1968).

Beermann, W., Bahr, G. F.: The submicroscopic structure of the Balbiani ring. Exptl. Cell Res. **6**, 195—201 (1964).

Berendes, H. D.: Induction and control of puffing. Ann. Embryol. Morph. Suppl. **1**, 153—164 (1969).

Clever, U.: Genaktivitäten in den Riesenchromosomen von *Chironomus tentans* und ihre Beziehungen zur Entwicklung. I. Genaktivierungen durch Ecdyson. Chromosoma (Berl.) **12**, 607—675 (1961).

d'Angelo, E. G.: Micrurgical studies on Chironomus salivary gland chromosomes. Biol. Bull. **90**, 71—87 (1946).

DeRobertis, E. D., Nowinski, W. W., Saez, F. A.: General Cytology, 3rd Ed., p. 251. Philadelphia (Penn.)-London: Saunders Comp. 1960.

Georgiev, G. P., Samarina, O. P.: Nuclear complexes containing informational RNA. Ann. Embryol. Morph. Suppl. **1**, 81—87 (1969).

Gruzdev, A. D., Belaja, A. N.: The influence of the concentration of hydrogen ions, tonicity and ionic strength of solutions on the size of polytene chromosomes (trans. from Russ.). Tsitologia **10**, 297—305 (1968a).

— Belaja, A. N.: On the linkage in polytene chromosomes (transl. from Russ.). Tsitologia **10**, 995—1001 (1968b).

Kabat, D.: Fibrous complexes of deoxyribonucleic acid with certain globular proteins. Role of divalent metal ions in the organization of nucleoprotein structures. Biochemistry **6**, 3443—3458 (1967).

Kroeger, H.: Zellphysiologische Mechanismen bei der Regulation von Genaktivitäten in den Riesenchromosomen von *Chironomus thummi*. Chromosoma (Berl.) **15**, 36—70 (1964).

— Potentialdifferenz und puff-Muster. Elektrophysiologische und cytologische Untersuchungen an den Speicheldrüsen von *Chironomus thummi*. Exp. Cell Res. **41**, 64—80 (1966).

Langendorf, H., Siebert, G., Lorenz, I., Hannover, R., Beyer, R.: Kationenverteilung in Zellkern und Cytoplasma der Rattenleber. Biochem. Z. **335**, 273—284 (1961).

LEZZI, M.: Die Wirkung von DNase auf isolierte Polytän-Chromosomen. Exp. Cell Res. **39**, 289—292 (1965).

— RNS- und Protein-Synthese in puffs isolierter Speicheldrüsen-Chromosomen von Chironomus. Chromosoma (Berl.) **21**, 72—88 (1967a).

— Cytochemische Untersuchungen an puffs isolierter Speicheldrüsen-Chromosomen von Chironomus. Chromosoma (Berl.) **21**, 89—108 (1967b).

— Spezifische Aktivitätssteigerung eines Balbianiringes durch Mg^{2+} in isolierten Zellkernen von Chironomus. Chromosoma (Berl.) **21**, 109—122 (1967c).

— Differential gene activation in isolated chromosomes. Int. Rev. Cytol. **29**, 127—168 (1970).

— GILBERT, L. I.: Control of gene activities in the polytene chromosomes of *Chironomus tentans* by ecdysone and juvenile hormone. Proc. nat. Acad. Sci. (Wash.) **64**, 498—503 (1969).

— — Differential effects of K^+ and Na^+ on specific bands of isolated polytene chromosomes of *Chironomus tentans*. J. Cell Sci. **6**, 615—628 (1970).

ROBERT, M.: Einfluß von Ionenstärke und pH auf die differentielle Dekondensation der Nukleoproteide isolierter Speicheldrüsen-Zellkerne und -Chromosomen von *Chironomus thummi*. Chromosoma (Berl.) **36**, 1—33 (1971).

STEVENS, B. J., SWIFT, H.: RNA transport from nucleus to cytoplasm in Chironomus salivary glands. J. Cell Biol. **31**, 55—77 (1966).

Replication in Polytene Chromosomes

George T. Rudkin

The Institute for Cancer Research
Philadelphia, Pennsylvania

I. Introduction

Replication of polytene chromosomes has been reviewed frequently in the last two decades, either as a facet of chromosome replication in general or in treatments restricted to polytene nuclei. The reader is referred especially to the following for further discussion and bibliographic references: Beermann (1962), Pelling (1969), Taylor (1969), Prescott (1970), Swift (1969), John and Lewis (1969), Wagner (1969), and Pavan and da Cunha (1969). Most of the work reviewed in the following has been carried out on salivary glands of the higher Diptera. The view is taken that their chromosomes are synthesized by the same basic metabolic machinery that is used in other tissues and that phenomena observed in them should provide insights into the nature of the replication process in all cells.

II. Development of Salivary Glands

Polytene chromosomes of multicellular organisms develop in cells which are destined to grow in size but not to divide. They result from chromosomes that are replicated but not separated (see Fig. 1). Most of those studied are in larval tissues which are histolysed during pupal metamorphosis, but some are in cells that persist to the imaginal stage, as in Malpighian tubules, or develop into polytene form at later stages, such as in foot-pad or nurse cells. They also occur in some plant cells and in some ciliates (Ammerman, 1964; Nagl, 1969). Most studies of replication have been carried out on salivary gland nuclei and these will be the main object of the following discussion. Since virtually all of the pertinent data concern DNA, the main treatment will be restricted to that compound.

Salivary glands arise as invaginations from the blastoderm near the middle of the embryonic period (8 h) in *Drosophila melanogaster* and develop into paired organs containing approximately 125 cells each by the end of that period. It appears likely that no cell division takes place after invagination begins (Sonnenblick, 1950). The replications that lead to the giant polytene cells in late larval stages have already begun by the time the larva hatches (Rudkin, 1964). Microspectrophotometric measurements of DNA contents assembled from the literature (see Fig. 2) suggest that the time to complete a cell cycle increases as the cells and nuclei enlarge. On the assumption that the ectodermal cells have a 2C value at invagination, less than

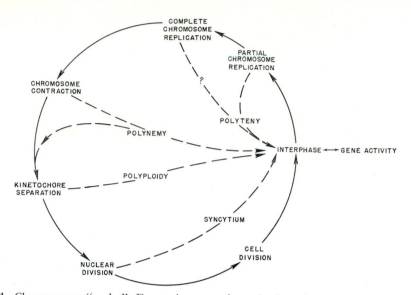

Fig. 1. Chromosome "cycles". Events in a complete mitotic cycle are arranged in a circle (solid line) and the curtailed cycles by which the amount of chromosomal material within a cell membrane is increased are indicated by the dashed "short cuts". The question mark indicates that although some of the genome remains unreplicated in some polytene nuclei, too few cases have been studied to make a general statement. Polynemic nuclei are thought to go through a few curtailed cycles, then through the complete cycle with polyvalent chromosomes (See Rudkin, 1965; Kaufman, 1965; Gay et al., 1970)

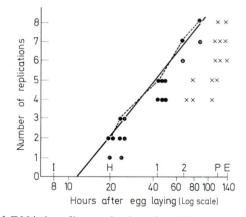

Fig. 2. Replication of DNA in salivary gland nuclei of larvae of *Drosophila melanogaster*. Ordinates show the number of replications (0, or no replications, represents a diploid nucleus). The abcissae represent, on a logarithmic scale, hours after egg deposition at 25° C. Developmental events are marked as follows: *I* invagination of the salivary gland plaque; *H* hatching from the egg; *1* and *2* the first and second larval molts; *P* puparium formation; *E* eversion of the imaginal discs. The dots are the data of Rudkin (1969), the crosses the data of Rodman (1967). The number of replications (ordinates) takes into account only the fraction of DNA that is actually replicated (see text). Dashed lines are drawn between the largest number of replications observed at each age. A solid line is drawn by eye through the points of maximum replication number and extrapolated to the abcissal axis

12 h is available to reach the 8C value observed in freshly hatched larvae. The calculated maximum average time of 6 h between those first doublings appears to increase logarithmically through much of the larval period, that is, the cell cycle period is roughly proportional to the DNA content of the nucleus. It is reasonable to believe that the S period itself also lengthens, though not necessarily proportionately; the proportion of cycle time taken up by periods corresponding to G1 and G2 is not known.

The correlation between changes in DNA synthesis and metamorphic events has led to the assumption that replication is under hormonal control (see KRISHNAKU-MARAN et al., 1967). DANIELI and RODINÒ (1967) report that the rate of incorporation of ^{3}H-TdR, measured by scintillation counting in salivary glands of *D. hydei*, drops precipitously at the time of the second larval molt and quickly resumes a high level at the beginning of the third instar. The low uptake was reflected in decreased numbers of labeled cells and fewer autoradiographic grains per labeled nucleus in parallel studies. DARROW and CLEVER (1970) observe the same phenomenon both in salivary gland and in malpighian tubule nuclei of *Chironomus tentans*. CROUSE (1968) found a stage in the last larval instar of *Sciara coprophila* at which virtually no replication could be detected in salivary gland nuclei 3—4 days after the last larval ecdysis; the molting process was not investigated. Comparable observations in *Rhynchosciara angelae* salivary glands (MATTINGLY and PARKER, 1968) were also not extended as far back as the third instar but the high frequency of labeled nuclei in the youngest larvae observed is similar to the burst of synthesis observed post-ecdysis in the other forms. Replication appears to stop in most polytene cells of Sciaridae near the middle of the fourth instar but resumes again in malpighian tubules and salivary glands during the periods of larval migration and cocoon formation. Still a third pattern, seen in polytene cells of the testis which incorporate ^{3}H-TdR during the whole period, shows that the control of replication is tissue specific and is not related to the polytene state. Conflicting observations that the administration of ecdysone derivatives induces synthesis in salivary gland chromosomes of Sciara (CROUSE, 1968) but not of Chironomus (DARROW and CLEVER, 1970) may find their explanation in the different stages at which the hormone was tested. CROUSE used post-ecdysis larvae in which replication was almost totally repressed and examined larvae the next day, while DARROW and CLEVER chose relatively older larvae in which the burst of synthesis in the last instar had already begun and waited two to three days to score them. While control is related in time to metamorphic events, and therefore to developmental changes in the levels of certain hormones, the direct response, if any, to the hormones is yet to be established.

The occurrence of discrete classes of DNA contents at all ages (WELSCH, 1957; RODMAN, 1967a; PETTIT, RASCH, 1967; RUDKIN, 1969) suggests that S periods within the larval instars are separated in time, a conclusion supported by autoradiographic evidence in Rhynchosciara (PAVAN, 1965b) and Chironomus (DARROW and CLEVER, 1970). The repression of replication at molting is most easily understood in terms of a block to initiation before ecdysis begins and its removal after the molting process has been completed. The prepupal change in Drosophila is followed by initiation in only a very few nuclei (RUDKIN and WOODS, 1959; RODMAN, 1968) but synthesis continues in some of them throughout the prepupal period. In all forms studied, those few nuclei still replicating at or shortly after the end of an instar appear to be in the terminal phases of S in keeping with the above interpretation. It follows that the chromosomes of the last larval instar in which patterns of synthesis have been studied do not come from a randomly replicating population of nuclei but from a population at least partially synchronized at the previous molt.

III. Intrachromosomal Patterns of DNA Synthesis

A. Ordered Replication

The long final period of DNA synthesis and the wealth of chromosomal detail in the giant chromosomes have made possible analyses of intrachromosomal patterns of synthesis on a much finer scale than can be carried out with dividing cells. The incorporation of radioactive thymidine followed by autoradiography reveals portions of chromosomes undergoing synthesis (Plaut, 1963) with a resolution in time that is higher the shorter the radioactive pulse. The absence of a real time marker and the partial asynchrony of the synthesis periods of the nuclei within a single gland or larva have called forth special techniques to supplement the analysis.

Order in the replication of small chromosome segments was discovered by Keyl and Pelling (1963) who were able to seriate about two dozen heavy bands in Chironomus with respect to which ones were labeled in single nuclei. The series progressed in an orderly manner from sub-sets of bands labeled when no others were to a sub-set labeled only when all others were. The authors interpreted their series as a cyclic permutation of the order in which synthesis began or ended in the bands they scored. The direction of the time vector was assessed from double labeling experiments in which an injection of ^{14}C-TdR was followed by one of ^{3}H-TdR. Since less of the chromosome was labeled by ^{3}H than by ^{14}C, they concluded that patterns with all of the bands labeled (since called the "continuous label" pattern) occurred earliest and that their series represented the order in which the individual bands completed synthesis. Their data further indicated that kinetochore regions were the last to complete replication. Although they could not exclude the possibility that they were in error by part of a replication cycle, i.e., that ^{14}C label was administered at the end of one S period and ^{3}H thymidine (6 h later) at the beginning of the next one, other considerations (see below) render that interpretation unlikely. Having thus shown that some bands continued replication after most others had finished, they made the hypothesis that none began synthesis significantly sooner than others. The continuous labeling pattern was assumed to represent the beginning and all "discontinuous" patterns later stages of the S period. The implication that a single signal initiates all replication units ("simultaneous" model) makes it important to examine the evidence closely.

The hypothesis has been tested most extensively in Drosophila salivary gland chromosomes where intensive analysis of the labeling patterns has been carried out over small (about 6—30 band) segments of appreciable fractions (about 5—15%) of the genome. The model predicts a specific kind of pattern distribution that can be represented as a matrix (see Fig. 3) in which the columns represent the scored segments arranged in rank order of the frequency with which they are labeled, the rows individual chromosomal (nuclear) patterns arranged in rank order of the number of segments labeled in the pattern. The theory predicts that the matrix will be

Pattern Frequency	A	B	C	D	E	F	G	H	I	J	K
1	A	B	C	D	E	F	G	H	I	J	K
1	—	B	C	D	E	F	G	H	I	J	K
1	—	—	C	D	E	F	G	H	I	J	K
4	—	—	—	D	E	F	G	H	I	J	K
3	—	—	—	—	E	F	G	H	I	J	K
4	—	—	—	—	—	F	G	H	I	J	K
9	—	—	—	—	—	—	G	H	I	J	K
1	—	—	—	—	—	—	G	—	I	J	K
8	—	—	—	—	—	—	—	H	I	J	K
1	—	—	—	—	—	—	G	—	—	J	K
1	—	—	—	—	—	—	—	H	—	J	K
6	—	—	—	—	—	—	—	—	I	J	K
1	—	—	—	—	—	—	—	—	I	—	K
12	—	—	—	—	—	—	—	—	—	J	K
11	—	—	—	—	—	—	—	—	—	—	K

Site label Frequency	1	2	3	7	10	14	25	32	39	52	64

Pattern Frequency											
1	—	—	—	—	—	—	—	—	I	—	K
1	—	—	—	—	—	—	G	—	—	J	K
1	—	—	—	—	—	—	G	—	I	J	K
1	A	B	C	D	E	F	G	H	I	J	K
1	—	B	C	D	E	F	G	H	I	J	K
1	—	—	C	D	E	F	G	H	I	J	K
4	—	—	—	D	E	F	G	H	I	J	K
3	—	—	—	—	E	F	G	H	I	J	K
4	—	—	—	—	—	F	G	H	I	J	K
9	—	—	—	—	—	—	G	H	I	J	K
8	—	—	—	—	—	—	—	H	I	J	K
6	—	—	—	—	—	—	—	—	I	J	K
1	—	—	—	—	—	—	—	H	—	J	K
12	—	—	—	—	—	—	—	—	—	J	K
11	—	—	—	—	—	—	—	—	—	—	K

Fig. 3. The data of NASH and BELL (1968) for twenty segments on the distal portion of chromosome 2R *(D. melanogaster)* in 7-day-old larvae (20° C) are arranged in matrices in which each row represents a chromosome labeling pattern and each column represents a set of segments which are all either labeled or unlabeled in a given chromosome. The presence of a letter indicates the presence of ^{3}H-thymidine label in an autoradiograph, the presence of a dash (—) indicates the absence of a label. The segments (columns) are arranged in order of the absolute frequency with which they were labeled (given between the matrices). In the upper matrix the patterns (rows) are arranged in decreasing order of the number of labeled segments within them; the absolute frequency with which they where observed is given along the left edges of the matrices. The patterns have been rearranged in the lower matrix to reduce the number of interruptions in the lettered (labeled) columns from six to two. There were seven segments in the frequency 3 column (C), two segments in the frequency 7 column (D) and one segment in each of the other columns. This figure illustrates some properties of data and of models, using a real data set that is small enough to be easily reproduced. On either model the synthesis of a segment is assumed to be uninterrupted in time, which would mean that a column would be composed of a run of uninterrupted letters and a run of uninterrupted dashes. The upper matrix incorporates the assumption that all segments begin DNA synthesis simultaneously in the top row and complete it in a definite order in successively lower rows ("simultaneous model"). The segment classes lettered A to F and K fit the model but those lettered G to J do not. The restriction that synthesis begins in all sites at one time has been removed in the lower matrix ("cascade model") and the observed patterns (rows) were rearranged so that all except two segment classes (H and I) show uninterrupted columns. Independent evidence suggests that the time vector would point downward in this model, also. Segments lettered I and K would have begun synthesis before others in the chromosome segment studied

divided into two triangular areas, in one of which all sites are labeled, in the other of which no sites are labeled. The number of "labeling patterns" would be equal to the number of ranks of site-labeling frequencies; there could be fewer patterns than sites if two or more sites were labeled with the same frequency (hence of the same rank) but there could not be more patterns than ranks. It turns out that the number of patterns exceeds the number of ranks in nearly every published data set, which means that the model cannot be accepted in its simplest form.

Two modifications have been proposed (after making allowances for the obvious technical errors that would creep into any extensive set of autoradiographic data). The first, proposed by HOWARD and PLAUT (1968), is that two or more sets of patterns exist in the population of chromosomes examined. LAKHOTIA and MUKHERJEE (1970) found differences between males and females which would explain some of the disagreement between the simultaneous model and the X chromosome data of HOWARD and PLAUT (1968) for mixed sexes. HOWARD and PLAUT proposed two sets of patterns, one representing the order of initiation of S and the other representing the order of termination of S with no way of telling which was which. Their interpretation indicates initiation in one or a few sites followed by others ("cascade model"). The following discussion will show that it has not yet been experimentally distinguished from the simultaneous model.

The second and most popular modification of the simultaneous model is the trivial one that postulates a stochastic element in the determination of the moment when a region completes its replication, or, more properly, when a given pulse will no longer result in autoradiographic label at the region. Probably the most suggestive evidence for that interpretation is that most of the deviations from the model occur near the "diagonal" boundary between the fields of labeled and unlabeled states, as in Fig. 3. In addition, it is usually not possible to arrange the observed patterns into only two error-free series, although the number of exceptions can be drastically reduced if only two series are assumed. Also, where pattern frequencies have been published, one can arrange that one series consists only of very rare patterns while the other includes all of the more frequent ones [see NASH and BELL (1968), and HOWARD and PLAUT (1968)]. However, the data fit the cascade model equally well. While the scarcity of exceptional patterns can be interpreted to mean that deviations from a fixed order of completion of synthesis are rare, it could also mean (1) that the time from first initiation to "continuous pattern" is short compared to the pulse period and to the time from the first to the last completion of synthesis in any region, or (2) that the data set was taken so late in the third instar that very few "initiation patterns" were encountered. Evidence for a waning frequency of initiation was found by NASH and BELL (1968) who observed nearly twice as many patterns in 6-day-old larvae as at 6 1/2 or 7 days (at 20° C): the frequency of "exceptional patterns" and the frequency of nuclei entering S decrease together (see also RODMAN, 1968).

Not only is the question still open as to whether initiation occurs in all segments nearly simultaneously, it is also not known what fraction of the nuclear S period is taken up by the various patterns. Since (as noted above) partial synchrony is imposed at the second molt and initiation virtually ceases by the end of the third instar, the scored nuclei cannot represent a random sample with respect to the nuclear cycle. Selection was intensified by restricting the sample to a few hours of the development period in all published data.

Relative labeling frequencies of individual segments, which would give the order of termination of synthesis on the simultaneous model, can be highly reproducible. Fig. 4 shows that the rank orders of the labeling frequencies of 20 sites at the distal end of chromosome 2R are quite similar from late third instar male larvae raised at

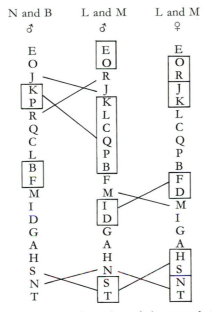

Fig. 4. Twenty segments on the distal portion of the second chromosome of *Drosophila melanogaster* are listed in increasing order of the relative frequencies with which they were found labeled by ^{3}H-TdR. The segments were serially lettered in their order on the chromosome, A being the most proximal, T the most distal (terminal) segment scored. Segments enclosed in a box were labeled with the same frequency. The first column is the combined data of NASH and BELL (1968), the second and third columns are from LAKHOTIA and MUKHERJEE (1970). The first two columns refer to male larvae, the third to female larvae. Instances in which rank orders of frequency differ are marked by lines between the columns. The figure illustrates the constancy in rank order of observed frequencies within a species, from which the existence of a reliable control mechanism is inferred. Reproducible rank order would mean for the "simultaneous" model that the order in which segments complete synthesis is constant and for the "cascade" model that the order of initiation of synthesis is constant also (see text). There is evidence that sex-dependent changes in rank order of labeling frequency are significant for autosomal segments as well as for the X chromosome segments in Fig. 5 (see text)

25° C and at 20° C in different laboratories and from different strains. The few changes in rank indicated by the data may be trivial, but some of them could represent real variations due to genetic or environmental disparities between the two experiments. Instances of genetic influence will be brought up later.

The interpretation of patterns is no less difficult for polytene chromosomes of other forms. Though the data are not as detailed as that for Drosophila, inferences have been drawn from bandwise studies of small portions of the genome and from coarse observations on whole chromosome complements. Thus in Chironomus species (Hägele, 1970) and in Sciara (Gabrusewycz-Garcia, 1964) complementary patterns are found which are interpreted as representing the beginning and ending of the S period, and continuously labeled chromosomes are assigned the intervening time. The assignment of late patterns is supported by independent evidence that such patterns do occur late in the last cycle; they include dense bands with a high content of DNA and regions situated near the ends of chromosome arms (kinetochore and telomere). The complementary patterns, typified by disperse regions such as puffs in Sciara and interband regions in both genera, are assumed to occur early. While the assignment of order to the three categories of pattern is probably correct, it is still an open question whether all segments start off at the same signal or enter S in a programmed sequence. A reasonable interpretation of the data would be that tritium incorporation is first detectable in the most disperse regions, possibly because it begins there and possibly because it can proceed more rapidly in disperse than in compacted chromatin. Where detailed analysis has been carried out on short segments (see Hägele, 1970), it is quite clear that the observed patterns cannot be arranged in a matrix that reveals an unequivocal order for cessation of synthesis, but it is evidently very difficult to acquire comparably detailed quantitative information known to represent the beginning of S. Darrow and Clever's (1970) observation that discontinuous patterns are least frequent early in the last instar indicates that if they occur first they must be of short duration.

A major drawback is ignorance of the absolute time differences involved. Were two segments to require on the average nearly equal replication times, then secular variations in the moment of initiation or in the time to label the local thymidine pools or in the local autoradiographic efficiency could introduce statistical deviations from the model, as mentioned above. If time intervals represented by pattern changes are large compared with S, then the model must be modified to accommodate them.

Polyteny develops in foot-pad nuclei of *Sarcophaga bullata* during the early pupal period (Whitten, 1964). Bultmann and Clever (1970) report that the fraction of labeled nuclei showing discontinuous patterns increases from about 66 to 100% between day 5 and days 6−8. Since no synthesis was detected after day 8, they conclude that, as in other forms, discontinuous patterns occur late. However, the details of the patterns were not studied nor were attempts to observe the beginning of S reported.

The simplest model compatible with all of the data is that synthesis begins in many, possibly all, disperse chromosome regions within a short period of time, then proceeds in all regions until the synthesis of the segment requiring the longest time has been completed. The observations are, however, equally satisfied by the postulate that initiation occurs first in one or a few sites and is taken up by others in a fairly definite, rapid sequence. The distinction has not been made experimentally and may have to await the identification of the chemical factor(s) involved.

B. Modification of Patterns

Reports of modifications of patterns of synthesis by duplication or deficiency of chromosomal segments must be interpreted with caution. The usual observation is that nuclei in which the segment with higher DNA content is labeled when its lesser homolog is not labeled occur more frequently than the contrary class of nuclei. It is then argued that the larger segment must require more time for its synthesis. However, the finite duration of the pulse label, the stochastic nature of autoradiographic data and the fact that the different nuclei in a sample are not synchronous for S can combine to produce just such observations even when the segments in question are made up of elements with coeval replication periods. For example, a triplication for a given segment exists in every scored nucleus as two paired homologs and an unpaired homolog. Assume that the replication periods are identical in all three homologs within a nucleus, an assumption supported by the observation of BARR, VALENCIA, and PLAUT (1968) that homologous segments on unpaired chromosomes in nuclei heterozygous for a rearrangement do not differ from one another in their relative labeling frequencies or positions in the labeling pattern matrix. Let the probability that the unpaired homolog will not exhibit label be p_0. The probability that the paired segment will not exhibit label is then p_0^2. Since $0 < p_0 < 1$ in any meaningful experiment, the sample will contain a higher frequency of unlabeled, unpaired than unlabeled, paired segments. Thus, nuclei with only the paired segment labeled would outnumber those with only the unpaired segment labeled, as observed. The expected frequencies of the four possible combinations can be predicted as functions of po. BENDER, BARR and OSTROWSKI (1971) have published data for a duplication of a segment of the X chromosome in salivary gland nuclei of female *D. melanogaster*, which fit the predicted frequencies. It may be concluded that the replication patterns are synchronous in all three homologs, contrary to the author's views. A contingency test applied to *non homologous* regions usually does not fit the coeval assumption. For example, the frequencies for sites I and J in Fig. 3 differ from expected (in a 2×2 table) at a probability level much less than 0.001.

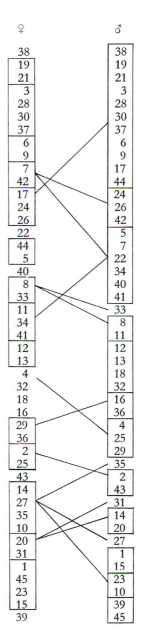

Fig. 5. Forty-five segments in the distal half of the X chromosome in *D. melanogaster* salivary gland nuclei are listed in order of increasing relative labeling frequency (^{3}H-TdR) reported by LAKHOTIA and MUKHERJEE (1970). The segments were numbered in order from the free end of the chromosome (1) to the most proximal scored segment (45). The figure illustrates the effect of sex on the relative labeling frequencies of segments of X chromosomes, to be compared with the lesser effect on an autosome seen in Fig. 4. For the "simultaneous" model, the differences would be interpreted as changes in the time required for replication of different segments, while for the "cascade" model, changes in order of completion of synthesis could not be distinguished from changes in order of initiation. (see legend to Fig. 4 and text)

The above result has particular relevance to the conclusion reached by Berendes (1966) and Lakhotia and Mukherjee (1970) that the X chromosome completes its replication relatively earlier in S in male larvae (where it is haploid) than in female larvae (where it is diploid and paired). The analysis is more difficult because the sex chromosomes must be compared with each other by reference to the paired autosomes within the respective nuclei. However, the original data require reexamination in the light of the above discussion.

Multiples of sets of complete replicating units, as in the X chromosomes and fragments noted above, can be assumed to include initiation sites. If the data did not fit the stochastic model, then a differential effect on time of initiation, duration of synthesis or both would be implied. Very short duplications within a chromosome which involve DNA resolved as a single band might, on the other hand, be produced by a mechanism that permitted the loss of initiation sites from duplicate copies. Such cases have been studied in subspecies of Chironomus where certain heavy bands contain multiplies of a minimum amount of DNA, multiples restricted to powers of 2 (Keyl, 1964). Keyl and Pelling (1963) found the denser, larger of two homologous bands more often labeled in nuclei of hybrids. Their interpretation in terms of extended duration of synthesis would imply that the duplications lacked initiation sites and were, therefore, replicated in tandem with an ancestral segment. Examination of the original data would be required to test the alternative hypothesis. Similar considerations would apply to duplications for the *white* locus studied in *D. melanogaster* males by Arcos-Teran and Beermann (1968).

Consistence with the statistical predictions would not, of course, establish the coeval hypothesis; it would remove the necessity to invoke other mechanisms to explain the results. Disagreement with predicted frequencies would exclude coeval replication periods but would not distinguish between differences in duration of replication periods and asynchrony in the timing of equal (or unequal) periods.

There are, however, instances in which changes in pattern have been detected in different genotypes. Fig. 5 reveals several instances in which pairs of X chromosome segments occur in reverse rank order of labeling frequency in male and female larvae. Other alterations are not directly indicated by the obvious changes in rank order: for example, segments 39 and 45 were always both labeled or both unlabeled in males, whereas segment 39 was sometimes the only one labeled in females. Such differences are not confined to the X chromosome, for Berendes (1966) noted a switch in the rank order of two sites on an autosome and similar changes are displayed for the second chromosome in Lakotia and Mukherjee's sample (Fig. 4). The combinations of bands in the kinetochore region labeled in the four chromosomes of Chironomus are not the same in intersubspecific hybrids as they are in the parent subspecies (Hägele, 1970). The different genetic environments must differentially affect some chromosome segments with respect to initiation sequence or to time required for replication or, perhaps, both. Locus-specific amplification, though not excluded, has not been observed in either genus.

The conclusion that has been drawn from the above observations, that duration of synthesis is positively correlated with DNA content of the band or region scored, has not been established for homologous regions. When non-homologous regions are compared, there is no correlation between DNA content and either labeling frequency or position in the sequence of patterns arranged for a best fit to the simultaneous model. It does appear to be true, however, that the regions in which DNA synthesis is completed last contain very compacted bands with a high DNA content. Variations in structure and arrangement of chromosome fibrils in different chromosome bands as seen in electron micrographs (SCHULTZ, 1965; SCHULTZ and ASHTON, 1971) suggest a possible reason for the poor correlation; the rate of synthesis may be influenced by the protein moiety to which variations in structure are attributed. On the other hand, more than one unit of replication within a band that appears to be "single" in the light microscope, or sequential synthesis through two or more bands from a single initiator locus would also serve to destroy the correlation. It should also be pointed out that there is no evidence that would indicate how closely initiation or replication is synchronized in the polytene strands of a single band. It is usually tacitly or overtly assumed that all begin simultaneously in one replicating segment, but asynchrony among them dependent on factors other than DNA content would equally effectively dissociate the two variables. Finally, for the cascade model in which initiation is effected region by region in a specific pattern, the relative order of cessation of synthesis would not be a function of duration of synthesis and the variance of the correlation would be increased according to the amount of time required to initiate all the replicating units.

C. The Replicating Units

No effective attempt has been made to determine the sizes of the replicating units in polytene tissues. The ultimate magnitude discernible with the light microscope is a chromomere or band, but the achievement of resolution to that level, very difficult autoradiographically, would at best set an upper limit. A dramatic example of the uncertainty inherent in the data is given by comparison of two analyses of the terminal three sections of chromosome 2R in *D. melanogaster* which HOWARD and PLAUT (1968) divided into 8 segments but in which NASH and BELL (1968) and LAKHOTIA and MUKHERJEE (1970) scored 14 segments. Essentially the same results are obtained for all sets of data if one scores a "HOWARD and PLAUT" segment as labeled when any one of its subsegments is labeled in the other data. Obviously, reducing the resolving power, as by scoring the chromosome in segments several bands long, does not imply that a unit of replication may extend through several bands.

Bands as the smallest resolvable subdivisions of chromosomes have not been differentiated from interband regions in a functional sense. Individual gene functions have been localized at specific bands which have been called "cytogenetic units" (RUDKIN, 1965) on the hypothesis that each band (plus interband) contains the DNA for a single Mendelian unit of heredity. There is also evidence that a band may be controlled as a unit of genetic activity and PELLING (1966) has extended the concept to propose that it is also a polytene unit of replication. The justification for the last hypothesis rests mainly on the independence of fairly large bands in the times at which they finish synthesis. However, no conclusive data exist with regard to whether a single initiation occurred within such a band (or an adjacent interband) or whether the replication process spread to it from a neighboring region. DNA contents of individual bands are worth examining with respect to the question.

The haploid genome of *Drosophila melanogaster* contains 0.18 picograms of DNA[1] of which 78% (0.14 pg) is represented in the banded polytene chromosomes (see Rudkin, 1965; 1969). Bridges figured slightly more than 5000 bands[2] from which is derived an average band value of 17×10^6 daltons per haploid strand. That corresponds to approximately 9 μm of double helical DNA. A similar average value for the haploid state has been arrived at for the banded regions in *D. hydei* chromosomes by Mulder, van Duijn and Gloor (1968) (0.16 pg) but those authors obtain a larger average band because Berendes (1963) counted about 2000 bands in *D. hydei* salivary gland chromosomes. Since Berendes (1970; see also Derksen and Berendes, 1970) counts fewer bands in *D. melanogaster*, the two average values are in good agreement with each other but uncertain by a factor of approximately two. Rudkin (1961), on the other hand, measured UV absorbance in a sample of 100 individual (Bridges) bands in polytene nuclei for which the degree of polyteny could be calculated. The haploid strand values ranged from 3×10^6 to approximately 365×10^6 with a mean value of 21×10^6 daltons. In terms of length, the range is from 1.7 to 200 μm at the chromatid level. Chironomus chromosome bands appear to have the same average haploid DNA content, within a factor of 3, as determined by Edstrom (1964) using microchemical procedures and band counts by Beermann (1952).

In the polytene chromosomes where the replicating units are scored, the 2000 chromatids, assuming proportionate synthesis[3], contain 2000 times as much DNA as a haploid strand or the equivalent of 3.4—400 mm. Assuming an S period of 8 h for a whole nucleus and allowing all of it for an individual band, one obtains calculated average minimum rates of 7—830 μm per min if all of the DNA of a polytene band were to be replicated in tandem sequence. Thus, larger bands would not have time to be replicated even at the high rate of 20—50μm/min reported for *E. coli* (Cairns, 1963) and no band would be completed at eukaryote rates estimated to be from 0.5—2 μm per min (Cairns, 1966; Hubermann and Riggs, 1968; Taylor, 1968). The conclusion that many, if not all, chromatids must be replicating simultaneously is not surprising. At the chromatid level, the calculated minimum rates for an 8-h-S-period turn out to range from 0.004—0.4 μ per min, well within measured ranges for other eukaryotes. A different picture is encountered, however, if one assumes the same organization into replicating units in early cleavage nuclei where inter-mitotic times are as short as 10 min in Drosophila: the calculated minimum rates come out to be 0.7—20 μ per min. Either the conditions at early cleavage are such as to permit replication rates approaching bacterial speeds or one is forced to the

---

1 Rudkin (1965) gives an integrated absorbance of 0.96 at 257 nm, which corresponds to 0.369 pg for diploid ganglion cells or 0.18 pg for the haploid value. Rasch, Barr and Rasch (1971) obtained the same haploid value for *D. melanogaster* sperm by measuring Feulgen dye bound to sperm nuclei and to nuclei with known DNA contents.

2 The number of bands (5059) is obtained by adding the numbers figured for the X chromosome, chromosomes 2R, 2L, 3R, 3L and 4 in Bridges (1938), Bridges (1939), Bridges (1942), Bridges (1941a), Bridges (1941b) and Bridges (1935), respectively.

3 Schultz and Rudkin (1960) fed ^{3}H-TdR in the first day of larval life, then raised larvae on non-radioactive medium to the third instar. Grain counts over short segments were not proportional to the known DNA contents of the segments. The ratios they observed are within expectations based on present knowledge of replicating units and cannot support their suggestion that disproportionate replication may have occurred.

conclusion that bands may contain more than one initiation site per chromatid, most of which are inactivated during embryogenesis. The high content of polyamines in very early embryos (DIONE, 1970), failure to stain for histones before the 10th cleavage division (KAUFMANN et al., 1962) and the appearance of chromocenters only after the syncytial cleavage stages (SCHULTZ, ASHTON, and MAHOWALD, 1967, cited in MAHOWALD, 1968) suggest that major alterations in the basic protein moiety of chromosomes may be associated with the switch down from the high initial rates.

The above discussion is based on the assumption that germline chromosomes are unineme in the species concerned. The long-standing controversy over that question (see, e.g., PRESCOTT, 1970) indicates that it must, for the time being, be answered experimentally for each organism under discussion. Genetic (e.g., MULLER, CARLSON and SCHALET, 1961) and more recently, chemical (see LAIRD, 1971) evidence supports the unineme hypothesis for some Drosophila species.

The haploid lengths calculated from the DNA contents of bands are in the same range as the lengths of replicating units estimated for mammalian cells in tissue culture (10−200 μ) by HUBERMANN and RIGGS (1968) and TAYLOR (1969). They are thus consistent with the view that a band may consist of one unit of replication at the chromatid level but they do not exclude more than one such unit in a large band nor more than one small band in such a unit.

The sizes of replicating units have been discussed with respect to the lengths of fragments obtained by spreading DNA on a LANGMUIR trough and measuring in electron micrographs. The observed lengths range from 1.7 to 116 microns for Drosophila depending on species and treatment (DERKSEN and BERENDES, 1970), and from 5−155 μ for Chironomus (WOLSTENHOLME, DAWID and RISTOW, 1968) depending on species. While these results are in comfortable agreement with the cytophotometric data above, the authors' conclusions with respect to the number of bands traversed by the isolated pieces depend on the additional assumptions that (1) the only places where DNA strands would break during preparative procedures correspond to interband regions and that (2) all such places would be broken during the preparative procedures. To be relevant to the present question, it would also be necessary to assume a single initiation site per isolated piece. The justification of those assumptions is sufficiently weak that the observations neither encourage nor discourage the view that a polytene chromomere contains only one kind of replicating unit.

D. Disproportionate Replication

Polytene nuclei offer unique opportunities to study the control of chromosomal DNA synthesis not yet fully exploited. The independence of replicating units is exaggerated in centric chromatin of Drosophila which is not replicated in polytene chromosomes (HEITZ, 1933; RUDKIN, 1965; 1969; MULDER, VAN DUIJN and GLOOR, 1968; GALL, COHEN and POLAN, 1971; DICKSON, BOYD and LAIRD, 1971) and in the DNA puffs of the Sciaridae in which more than one round of replication can occur during a single nuclear S period (RUDKIN and CORLETTE, 1957; CROUSE and KEYL, 1968). In both cases, initiation is subject to region- specific control, on the one hand repressed and on the other hand activated.

Centric regions in Drosophila (including most of the Y chromosome) typically remain in a compacted state through interphase in dividing cells. They continue

DNA synthesis after the remainder of the nuclear DNA has finished in these as in other eukaryotes (Barrigozi et al., 1966; Halfer et al., 1969). Such "late replicating" regions frequently initiate synthesis relatively late in the nuclear S period in other forms, but that aspect has not been studied in Drosophila. It has been suggested that in salivary gland polytene nuclei, they "lag" behind the rest of the nuclear DNA for the whole of larval development in salivary glands (Swift, 1965), presumably for only a few replications in ganglion cells (Berendes and Keyl, 1967; Mulder, van Duijn and Gloor, 1968). Fox has reported a similar phenomenon in Locustidae (1970) and in Dermestidae (1971). The regions do appear to be subject to separate control systems. Centric location has suggested involvement of the kinetochore region, but in Drosophila, failure of kinetochore division is as likely to be a result of, as a cause of, suppression of replication in the neighboring chromatin. Nucleolar DNA, which is replicated in salivary glands (see Nash and Plaut, 1965) is located in the center of non-replicated segments on the X chromosome and the short arm of the Y chromosome in *D. melanogaster* (see Cooper, 1959), implying at least two non-replicating segments separated from the centromere by replicating ones. Consideration of the DNA contents of the regions involved shows that there must be more than two units. The centric regions of the X alone contain 3.7% of nuclear DNA (Rudkin, 1965) corresponding to 4.2×10^9 daltons, or about 2.3 mm at the haploid strand level, too much to be replicated in one S period at any reasonable rate. The foregoing considerations suggest that there would have to be of the order of 10 or more replicating centric X chromosome units in addition to others in the Y and in the autosomes, all subject to at least one common control factor.

Recent molecular hybridization data of Hennig and Meer (1971) suggest that *D. hydei* salivary gland nuclei contain fewer ribosomal DNA cistrons than would be predicted from the degree of polyteny. Intranucleolar DNA can, however, be labeled with ³HTdR throughout larval life in that species (Berendes, personal communication). It is not known that all intranucleolar DNA is ribosomal and, indeed, the argument can be made that the compact chromatin seen there (Barr and Plaut, 1966; Olvera, 1969) is inactive (Schultz, 1965) and, therefore, more likely to be derived from chromatin adjacent to the rDNA cistrons themselves. However, if it is assumed that at least some of the incorporation is into rDNA then Hennig and Meer's (1971) data would imply that not all of the cistrons were replicated in every polytene S period during development. Indeed, 150 copies of 38S precursor rRNA (about 3.3×10^6 Daltons) would be about 0.5 μm long and require a speed of about 1 μm per minute from one starting point in 8 h. However, the potential for varying the number of ribosomal cistrons per "genome" found in Drosophila (Henderson and Ritossa, 1970; Tartof, 1971) is too poorly understood to permit useful conclusions to be drawn regarding the number of replicating units within the nucleolar organizing region. The essential point in the present context is that at least one such unit separate from adjacent, non-replicating units must be postulated.

Supernumerary chromosomes resembling the Y chromosome remain compact and are apparently not replicated in polytene nuclei of *Phryne cincta*. The phenomenon is easily demonstrated cytologically when the supernumeraries are not fused with other chromosomal elements (Wolf, 1961) even though DNA measurements have not been reported. Dimorphic X chromosomes in the same species differ in length by a factor of two at meiosis but are of nearly equal length and are somatically paired in salivary gland nuclei. The member of the pair that is longer at metaphase has a block or blocks of compacted chromatin which, from quite convincing cytological evidence, have completed fewer replication cycles than has the rest of the chromosome (WOLF, 1968a). Evidently, all of the "extra" chromosomal material is subject to a common mechanism for the suppression of DNA synthesis.

The recent demonstration that proximal regions are especially rich in repetitive sequences and are the source of some satellite DNA's (PARDUE and GALL, 1970; JONES and ROBERTSON, 1970; PARDUE et al., 1970; HENNIG, HENNIG and STEIN, 1970; ECKHARDT and GALL, 1971; GALL, COHEN and POLAN, 1971) suggest base sequence as source of specificity for the control. The typically compact structure by which the regions are cytologically recognized in dividing cells implies that their DNA either interacts with a chromosomal protein in a characteristic, specific way, or that there are particular species of proteins with a high affinity for the sequences they contain. Speculations as to the controlling role of such proteins, or as to how replication would be varied from cell type to cell type in a particular set of regions are too numerous to be profitable at this time.

The so-called "DNA puffs" of the Sciaridae also appear to be under a separable control. Their replicating units, turned on and off along with the rest of the genome throughout the larval stages (PAVAN, 1965a), go through several cycles during and after the last nuclear S period (FICQ and PAVAN, 1958; RUDKIN and CORLETTE, 1957; CROUSE and KEYL, 1968; SWIFT, 1962; MATTINGLY and PARKER, 1968). DNA synthesized during the "extra" replications remains at the sites of the puffs; it is not of nucleolar origin (PARDUE et al., 1970; MENEGHINI et al., 1971). The replicating units involved appear to complete full rounds of replication as judged from the observation that the DNA contents of the puff sites form logarithmic series to the base 2. Ligation experiments of AMABIS and CABRAL (1970) show that initiation is dependent on viable tissue in the region of the larval brain, thus placing them in the category of "ecdysone puffs", known in other Diptera only as the RNA type: removal of the brain region with its associated endocrine organ prevented the formation of all puffs. (It is not excluded, of course, that a different hormone secreted simultaneously with ecdysone activates the DNA puffs). The DNA puffs can be selectively suppressed by interdicting the synthesis of DNA after a critical stage in larval development. SAVAIA, LAICINE and ALVES (1971) reduced the incorporation of ^{14}CTdR into salivary glands of *Bradysia hygida* (Sciaridae) to about one eight of its normal value by the injection of hydroxyurea. The suppression persisted for a period of about 20 h when induced at or after the critical period; were pre-injection levels never resumed. Before the critical period, recovery was more rapid and returned to normal levels, suggesting that synthesis of the riboside-diphosphate reductase, specifically inhibited by the drug (YOUNG, SCHOCHETMAN, and KARNOFSKY, 1967; ROSENKRANZ and CARR, 1970; SKOOG and NORDENSKJÖLD, 1971), proceeded at a much higher rate before than after the critical time. Concomitant cytological and autoradiographic studies (^{3}HTdR) showed normal puff development when injection before the critical period permitted recovery of DNA synthesis; suppression of puff development after later treatments depended on the time of injection. Given at the critical period, hydroxyurea prevented DNA puff formation and synthesis in other chromosome regions; given at later times, the puff development was interrupted at later stages and the ³HTdR incorporation into chromosomes reduced. Neither the appearance nor the ³H-uridine uptake of RNA puffs was affected by the inhibitor.

The clear result is that DNA synthesis is required for the development and maintenance of DNA puffs but not of RNA puffs, a result consistent with the view that hydroxyurea blocks the action of ribonucleoside-diphosphate reductase and so prevents the synthesis of DNA precursors without interfering with RNA metabolism. Low-level post-treatment labeling is understandable, for thymidine bypasses the biosynthetic block. There is no evidence for differential effects on DNA synthesis in the puff and non-puff regions of the nucleus. However, the "critical period" would seem to have physiological significance beyond the DNA puffs in the salivary gland chromosomes. Given at that, but no other, time hydroxyurea induces anomalies of adult development in a number of organs. Swindlehurst, Berry nad Firshein (1971) allude to comparable sensitive periods in silk moth larvae, different for different organs, correlated with extensive DNA synthesis in the affected developing organ. Thus the correlation between DNA puff suppression and morphological defects in the imago would seem to be a fortuitous one, as the authors suggest.

There are other DNA-containing bodies that are seen adjacent to chromosomal sites different from, or in addition to, the site of the main nucleolar organizer. It appears in some cases as though the extra-chromosomal DNA were disengaged from the chromosome at the site it is near and to which it often remains connected by a thread (that may be broken during preparation) (Whitten, 1965; da Cunha et al., 1969; Keyl and Hägele, 1966; Henderson, 1967). The obvious analogy with oocyte amplification of ribosomal (nucleolar) cistrons has been drawn (Pavan and da Cunha, 1969b). Pardue et al. (1970) did find ribosomal RNA hybridized with so-called "micronucleoli" (Gabrusewycz-Garcia and Kleinfeld, 1966) in the nuclei of Sciarid larvae. Nucleolar origin, so established, indicates that the bodies were probably derived not from the bands to which they adhered but from the nucleolus, which does regress during the last instar (Breuer and Pavan, 1955; Mattingly and Parker, 1968) and ceases to produce ribosomal RNA (Lara and Hollaender, 1967; Armelin, Meneghini and Lara, 1969). Hägele's (1970) view that DNA is continuously synthesized and shed from a locus in Chironomus may have a similar explanation. The origin of extra-chromosomal DNA bodies must be established by methods independent of morphology; so far, only nucleolar DNA has been identified.

Retention of the "extra" DNA on the chromosome sites in the Sciaridae (see Pavan, 1965a) and the under-replication of centric regions in Drosophila pose problems in chromosome structure. In both instances, adjacent replicating units undergo different numbers of replications, yet the physical continuity of the chromosome is maintained. Germ-line occurrences of locally doubled DNA contents are readily interpreted as linear arrays of duplicated segments, an evolutionary mechanism postulated long ago (Bridges, 1935; Muller, 1951) and recently applied to Chironomus by Keyl (1966). Mechanisms proposed to explain genetic recombination can be adduced (Schultz, 1965), tailored to the specific chromosome regions involved in the puffs and operating in all the homologous chromatids of the puff region. The recent discovery of RNA-dependent DNA polymerases (Temin and Mizutani, 1970) in normal cells (Penner, Cohen, and Loeb, 1971) opens a new area of speculation and investigation in the light of the presence of RNA (Pavan,

1965b; Ficq, Pavan and Brachet, 1958; Sauaia, Laicine and Alves, 1971) in the exceptional regions.

The Drosophila salivary gland chromosomes require different interpretations. That the linear structure of the chromosomes is weakened at the transition from replicated to non-replicated regions is suggested by the ease with which the X chromosome is broken at the site of the nucleolar organizer and the frequency with which the arms of mediocentric chromosomes are separated from one another in squash preparations. Detailed three-dimensional studies of polytene chromocentral regions following methods begun by Ashton and Schultz (1964, 1971) for imaginal cells should resolve the topological relationships in the transition zone between replicated and non-replicated segments.

IV. RNA and Proteins

Though RNA is universally found on polytene chromosomes and is synthesized on or at them, its role in replication would appear to be an indirect one. The relation between transcription and replication, particularly whether or not the two processes are mutually exclusive, is difficult to discover. RNAase-sensitive incorporation of RNA precursors is usually found in all nuclei of an organ, which implies that it is carried out in those chromosomes which are also being replicated. However, when thousands of bands, each with hundreds or thousands of chromatids are involved, the resolving power of methods so far utilized is insufficient to identify the molecules engaged in each activity (see Ritossa, 1964).

The synthesis of RNA is probably not essential to the usual replication process which can go on for long periods in third instar Chironomus larvae in the presence of actinomycin D. However, there is a requirement for RNA synthesized near the beginning of the instar because given then, the drug is effective in preventing or reducing subsequent incorporation of ^{3}H-TdR. Since cycloheximid is effective only at the same period of development, Darrow and Clever (1969) conclude that just after the last larval molt RNA is transcribed and immediately translated into protein that is essential for replication throughout the last instar. The normal ebbing of replicative activity as puparium formation is approached is then interpretable in terms of the exhaustion of a supply of essential materials built up after the last molt. Such questions as the function of the substances involved and their tissue of origin remain unanswered at present. The production of special substances may be related to the so-called "seventy-hour change" in Drosophila melanogaster larvae (Beadle, Tatum and Clancy, 1938) which occurs at about the time of the last larval molt. The "change" is such that younger larvae die if removed from an external source of food while older ones can complete development even if starved of all nutrient (except air and water). The mobilization of substances essential to chromosome replication may be a part of a special adaptation, with obvious survial value, peculiar to a stage of insect development, and of special interest as an experimental system.

On the other hand, protein synthesized during the last instar does appear on polytene chromosomes. Ficq, Pavan and Brachet (1958), Pettit and Rasch (1966) and Cave (1968) have demonstrated the incorporation of, respectively, ^{14}C phenyl-alanine, ^{3}H-histidine and ^{3}H-tryptophane into non-histone chromosomal proteins. By studying only chromosomes "continuously" labeled by ^{14}C-TdR applied simultaneously with the amino acid, Cave (1968) showed that both non-histone- and histone-type proteins (^{3}H-lysine) were synthesized and that they appeared on chromosomes during the replication cycle in Chironomus. If the assumption that cycloheximid prevented histone synthesis in Darrow and Clever's (1969) experiments is correct, then it would follow that histone synthesis need not accompany DNA replication, even though it does so in normal development: histones may be among the proteins "stored" at early third instar. Bloch and Teng (1969) suggest normal non-coupling of DNA and histone synthesis in spermatocytes of Rhenia. Wanka, Moors and Krijzer (1972) have partially uncoupled DNA and protein synthesis in Chlorella by the use of hydroxyurea and cycloheximid; proteins accumulated while the synthesis of DNA was blocked could support that synthesis after the block was removed, but only to the completion of the current round of replication. Initiation of S was still dependent on new protein synthesis. Thus, the storage hypothesis is unattractive and it may be useful to examine the histone/DNA ratio in the treated larvae, normally constant as measured by the alkaline fast green and Feulgen reactions (Alfert and Geschwind, 1953; Swift, 1964; Gorovsky and Woodard, 1961). Others have observed that, though puffs may be reduced, the banding patterns and morphology appear normal (Laufer, Nakase, and Vanderberg, 1964; Clever, 1967).

Changes in the amount of chromosomal protein associated with replication of DNA (Rudkin, 1964) and chromosomal localization of labeled amino acids with chromosomes in autoradiography such as noted above are, in general, open to various interpretations. Methods for more definitive demonstrations of function and better characterization of the protein species detected are required. Of special interest would be unequivocal experiments testing for intra-nuclear synthesizing systems (Ristow and Arends, 1967; Helmsing, 1970) and for their coordination, if any, with replication and transcription.

V. Controls of Replication

Most of what is known about replication in polytene nuclei is in the form of description from which the existence of controlling mechanisms can be inferred. Control may be postulated at at least three levels: one at the induction of, and possibly in the maintenance of, the polytene state; another responsible for maintaining complete, non-overlapping nuclear rounds of synthesis; and a third associated with single replicating units or specific sub-sets of units within the genome. A unitary hypothesis would interpret each level of control in terms of modifications of the co-ordinate synthesis typical of mitotic cells. At the present state of our knowledge, even that point of view tends to raise more questions than it answers.

We have seen that the chromosomes are made up of at least hundreds, and probably thousands, of replicating units which implies as many polymerases. Are the enzymes all the same, all different, or do they exist in "families" with similar properties? How are adjacent units joined to one another? These involve well-rehearsed questions of chromosome structure as yet unresolved but possibly mimicked in the patterns of control. It is clear that response to a signal to begin replication requires a responsive state in the replicating unit itself, and the specificity may be thought to lie either in the stimulus or on the chromosome. Thus, the reason for the failure of centric regions to respond in Drosophila salivary gland chromosomes may as easily be attributed to the absence of a particular initiator substance as to an inert state of the regions, brought on, for example, by the binding of a particular protein to the repeated sequences located there. In fact, the frequent observation that centric regions begin as well as cease replication later than other regions (see LIMA-DE-FARIA, 1969, for review) can also be interpreted both ways, and in either case only a minor modification is necessary to explain the polytene situation. The protein repressor theory has been put to a test by comparing the histones extracted from polytene nuclei with those from imaginal tissues of *Drosophila melanogaster*. Where the theory would predict that the salivary gland nuclei are deficient in a specific component, in fact no difference, qualitative or quantitative, was found by means of gel electrophoresis (COHEN and GOTCHEL, 1971). The negative result does not, however, exclude the hypothesis.

The necessity for initiation factor(s) outside of the chromosome is suggested by continued replication in glands transplanted to adult abdomens (HADORN, RUCH and STAUB, 1964; BERENDES and HOLT, 1965; STAUB, 1969), and by mutations which permit "extra" replications to proceed during extended larval periods leading eventually to complete development (RODMAN, 1967b; see also BRIDGES, 1935). A lethal mutation by which the larval period is extended but DNA synthesis in salivary glands stopped early in the third instar (WELCH and RESCH, 1968) is known to be deficient in hormone(s) necessary for metamorphosis (HADORN, 1937), and those hormones have already been implicated in chromosome replication in other systems (see above). Again, critical experiments to distinguish between the presence of an initiator and the absence of an inhibitor at the level of the DNA-polymerase complex have not been reported. Systems in which these questions may be profitably studied are found in parasitic infections of Rhynchosciara, which induce extra replications in polytene cells of several tissues (ROBERTS, KIMBALL and PAVAN, 1967; see also PAVAN and DA CUNHA, 1968; 1969b, for reviews), and in *Phryne cincta* where the relative size of the polytene X chromosome (DNA content has not been reported) is temperature sensitive (WOLF, 1957; 1968b).

Site-specific controls are also open to various interpretations. Controlling loci, separate from the affected locus, are suggested by the cytological manifestation of position-effect variegation studied by CASPERSSON and SCHULTZ (1938): chromosome bands placed near centric regions contained more DNA than did their homologs in the original chromosome sequence. An analysis of the structure of the affected regions with respect to the number and placement of replicating units is required for a proper interpretation of the spreading effect, but the possibility for the control of synthesis in a replicating unit by neighboring chromosome material is established (see SCHULTZ, 1947; 1965). Its application to ribosomal cistrons in salivary glands of Drosophila and in oocyte nuclei as well as to DNA puffs is obvious: DNA sequences governing the replication of others would provide a satisfying explanation of many of the phenomena observed in polytene and other nuclei (see TARTOF, 1971). The modulation of the activity of such controlling genes in different tissues would be accommodated by genetic variations among them, and their activity, insofar as close proximity in the linear structure of the chromosome is essential, may not require transcription.

Controlling functions in transcription and replication have been proposed for repetitive DNA sequences (see, e.g., BRITTEN and KOHNE, 1968; and WALKER,

Flamm and McLaren, 1969). Dickson, Boyd and Laird, 1971) find at least 5% of the DNA in polytene chromosomes of *D. hydei* to be in repeated sequences, and *in situ* hybridization studies show that repetitive DNA is distributed throughout salivary gland chromosomes (Pardue et al., 1970; Jones and Robertson, 1970; Rae, 1970; Hennig, Hennig and Stein, 1970; Gall, Cohen and Polan, 1971). Its function and relation to chromosome structure are still hypothetical.

VI. Concluding Remarks

Once the stricture is removed, that all germ-line DNA must be replicated for the preservation of a species, the classic question of how the information is distributed to somatic cells during a life cycle becomes important. There are obvious advantages to be gained by the elimination of DNA from cells in which it will never be used, and different advantages to the retention of information useful in sub-optimal or hostile environments but otherwise unnecessary for specific function in the course of normal development. Given that the chromosomes of somatic cells are descended from germ-line chromosomes, it follows that their information content could be adjusted by the modulation of mechanisms applicable to the replication of all chromosomal material and at the same time be subject to whatever limitations those mechanisms impose. It would seem that the possible variations are extremely broad, ranging from the elimination of whole chromosomes or chromosome sets, through the selective repression of replication in portions of chromosomes, to the amplification of selected DNA sequences. A major restriction imposed by a finite rate of synthesis necessitates the simultaneous replication of different portions of chromosomes but the portions so determined introduce the possibility of selective control. The polytene chromosomes indicate that the individual units are at least small enough to be replicated at known rates of synthesis, that they vary widely in DNA content and in information content, and that they can, in fact, be subject to separate control in the extent of their replication. The evolutionary advantages of polytene nuclei probably include the possibility for the generation of genic unbalance within chromosomes, suppressing the replication of information of no use in certain short-lived somatic cells, and so releasing precursors for the build-up of that DNA which is required. It would appear that multistranded chromosomes (paired or not) offer a solution to the problem that has not been achieved in evolution for chromosomes that must eventually divide (as in polynemic or polyploid nuclei).

The specific mechanisms involved are still obscure. New methods for their study have only begun to be exploited. Of special interest will be the extension of methods for the detection of substances and processes *in situ*, represented at the present time mainly by staining procedures, autoradiography and microspectrophotometry. *In situ* hybridization of DNA molecules (Gall and Pardue, 1969; John, Birnstiel and Jones, 1969) and an attempt to detect single-stranded DNA by the action of polamerases (von Borstel, Prescott and Bollum, 1966; von Borstel, Miller and Bollum, 1969) represent rewarding and, so far, unrewarding attempts, respectively. Techniques for the isolation of polytene cells from insects (Boyd, Berendes and Boyd, 1968; Cohen and Gotchel, 1971) require the prior separation of organs, a difficult task on a large scale, and one that would not be necessary in ciliates (Ammerman, 1971) if mass cultures can be successfully raised. Biochemical studies can be expected to elucidate pathways by which synthesis and activity of chromosomes are carried out and to provide a stable base for the design and interpretation of cytochemical experiments.

The patterns of replication are not likely to reveal the order of initiation in the various replicating units, unless a satisfactory method is discovered for timing the beginning of the S period in highly polytene nuclei. Isolation of DNA polymerase(s) and the determination of the chemical parameters within which they can function are to be expected. Of particular interest will be the investigation of repair mechanisms and their relation to the normal replication process. The characterization of other chromosomal proteins, their relation to the DNA's of different types of nuclei, changes in their proportion during nuclear growth and differentiation and their metabolic properties will have special value in polytene chromosomes where localization at separate parts of the genome will be possible. Possibly the most exciting hope, not yet in sight, is to discover the conditions necessary to induce the polytene state and to make them effective in the cells of organisms in which polyteny does not occur naturally.

Notes Added in Proof

CALLAN (CALLAN, H. G., Proc. Roy. Soc., ser. B, in press, 1972) has adduced evidence for different numbers of replicating units in nuclei of amphibian cells at different times of development, an observation relevant to the discussion of replicating units on page 69 ff.

RASCH (RASCH, E. M.: DNA cytophotometry of salivary gland nuclei and other tissue systems in Dipteran larvae. In: WIED, G. L., BAHR, G. F., Eds.: Introduction to Quantitative Cytochemistry- II, pp. 357–397, New York and London: Academic Press, 1970) finds that the DNA contents of polytene nuclei in *D. melanogaster*, *D. virilis*, *Ch. thummi* and *S. coprophila* fit successive doublings of the corresponding diploid DNA values (for *S. coprophila*, preceeding the appearance of DNA puffs). Since the evidence for "under-replication" of centric regions in Drosophila polytene cells (see page 71ff.) comes from such a wide variety of independent sources, the reason for the discrepancy should be sought at the methodological level (see also RODMAN, 1967a, and RUDKIN, 1969).

Acknowledgement

This work was supported by National Institutes of Healt grants CA-01613, CA-06927 and RR-05539; National Science Foundation grant GB-3525; and by an appropriation from the Commonwealth of Pennsylvania.

Abbreviations Used

TdR	Thymidine deoxyriboside.
S	The period during a cell cycle when DNA is being synthesized.
G1, G2	The periods between S and prophase and between telophase and S, respectively, in dividing cells.
^{3}H-X, ^{14}C-X	Compound "X" labeled with the radioactive isotope tritium or ^{14}Carbon.

References

Alfert, M., Geschwind, I. I.: A selective staining method for the basic proteins of cell nuclei. Proc. nat. Acad. Sci. (Wash.) **39**, 991—999 (1953).

Amabis, J. M., Cabral, D.: RNA and DNA puffs in polytene chromosomes of *Rhynchociara:* inhibition by extirpation of prothorax. Science **169**, 692—694 (1970).

Ammermann, D.: Morphology and development of the macronuclei of the ciliates *Stylonychia mytilus* and *Euplotes aediculatus*. Chromosoma (Berl.) **33**, 209—238 (1971).

Ashton, F. T., Schultz, J.: Stereoscopic analysis of the fine structure of chromosomes in diploid Drosophila nuclei. Cell Biol. **23**, 7 A (1964).

— — The three dimensional fine structure of chromosomes in a prophase *Drosophila* nucleus. Chromosoma (Berl.) **35**, 383—392 (1971).

Arcos-Teran, L., Beermann, W.: Changes of DNA replication behavior associated with intragenic changes of the *white* region in *Drosophila melanogaster*. Chromosoma (Berl.) **25**, 377—391 (1968).

Armelin, H. A., Meneghini, R., Lara, F. J. S.: Patterns of ribonucleic acid synthesis in salivary glands of *Rhynchosciara angelae* larvae during development. Genetics Suppl. **61**, 351—360 (1969).

Barigozzi, C., Dolfini, S., Fraccaro, M., Raimondi, G. R., Tiepolo, L.: *In vitro* study of the DNA replication patterns of somatic chromosomes of *Drosophila melanogaster*. Exp. Cell Res. **43**, 231—234 (1966).

Barr, J., Plaut, W.: Comparative morphology of nucleolar DNA of Drosophila. J. Cell Biol. **31**, C17—C22 (1966).

— Valencia, J. I., Plaut, W.: On temporal autonomy of DNA replication in a chromosome translocation. J. Cell Biol. **39**, 8a (1968).

Beadle, G. W., Tatum, E. L., Clancy, C. W.: Food level in relation to rate of development and eye pigmentation in *Drosophila melanogaster*. Biol. Bull. **75**, 447—462 (1938).

Beermann, W.: Chromomerenkonstanz und spezifische Modifikationen der chromosomenstruktur in der Entwicklung und Organdifferenzierung von *Chironomus tentans*. Chromosoma (Berl.) **5**, 139—198 (1952).

— Riesenchromosomen. Protoplasmatologia, Handbuch der Protoplasmaforschung, Bd. 6. Wien: Springer 1962.

Bender, H. A., Barr, H. J., Ostrowski, R. S.: Asynchronus DNA synthesis in a duplicated chromosomal region of *Drosophila melanogaster*. Nature New Biol. **231**, 217—219 (1971).

Berendes, H. D.: The salivary gland chromosomes of *Drosophila hydei* Sturtevant. Chromosoma **14**, 195—206 (1963).

— Differential replication of male and female X-chromosomes in Drosophila. Chromosome **20**, 32—43 (1966).

— Keyl, H. G.: Distribution of DNA in heterochromatin and euchromatin of polytene nuclei of *Drosophila hydei*. Genetics **57**, 1—13 (1967).

— Polytene chromosome structure at the submicroscopic level. I. A map of region X, 1—4E of *Drosophila melanogaster*. Chromosoma **29**, 118—130 (1970).

— Holt, Th. K. H.: Differentiation of transplanted larval salivary glands of *Drosophila hydei* in adults of the same species. J. exp. Zool. **160**, 299—318 (1965).

Bloch, D. P., Teng, C.: The synthesis of deoxyribonucleic acid and nuclear histone of the X chromosome of the *Rehnia spinosus* spermatocyte. J. Cell Sci. **5**, 321—332 (1969).

Boyd, J. B., Berendes, H. D., Boyd, H.: Mass preparation of nuclei from the larval salivary glands of *Drosophila hydei*. J. Cell Biol. **38**, 369—376 (1968).

Breuer, M. E., Pavan, C.: Behavior of polytene chromosomes of *Rhynchosciara angelae* at different stages of larval development. Chromosoma (Berl.) **7**, 371—386 (1955).

Bridges, C. B.: Salivary chromosome maps; with a key to the banding of the chromosomes of *Drosophila melanogaster*. J. Heredity **26**, 60—64 (1935).

— A revised map of the salivary gland X-chromosome of *Drosophila melanogaster*. J. Heredity **29**, 11—13 (1938).

— Bridges, P. N.: A revised map of the right limb of the second chromosome of *Drosophila melanogaster*. J. Heredity **30**, 475—476 (1939).

BRIDGES, P. N.: A revision of the salivary gland 3R-chromosome map of *Drosophila melanogaster*. J. Heredity **32**, 209—300 (1941a).
— A revised map of the left limb of the third chromosome of *Drosophila melanogaster*. J. Heredity **32**, 64—65 (1941b).
— A new map of the salivary gland 2L-chromosome of *Drosophila melanogaster*. J. Heredity **33**, 403—408 (1942).
BRITTEN, R. J., KOHNE, D. E.: Repeated sequence in DNA: hundreds of thousands of copies of DNA sequences have been incorporated into the genomes of higher organisms. Science **161**, 529—540 (1968).
BULTMANN, H., CLEVER, U.: Chromosomal control of foot pad development in *Sarcophaga bullata*. I. The puffing pattern. Chromosoma **28**, 120—135 (1969).
CAIRNS, J.: The chromosome of *Escherichia coli*. Cold Spring Harb. Symp. quant. Biol. **28**, 43—46 (1963).
— Autoradiography of HeLa cell DNA. J. molec. Biol. **15**, 372—373 (1966).
CASPERSSON, T., SCHULTZ, J.: Nucleic acid metabolism of the chromosomes in relation to gene reproduction. Nature (Lond.) **142**, 294—295 (1938).
CAVE, M. D.: Chromosome replication and synthesis of non-histone proteins in giant polytene chromosomes. Chromosoma **25**, 392—401 (1968).
CLEVER, U.: Control of chromosome puffing. In: GOLDSTEIN, L., (Ed.): The Control of Nuclear Activity, pp. 161—186. Englewood Cliffs, N. J.: Prentice-Hall 1967.
COHEN, L. H., GOTCHEL, B. V.: Histones of polytene and nonpolytene nuclei of *Drosophila melanogaster*. J. biol. Chem. **246**, 1841—1848 (1971).
COOPER, K. W.: Cytogenetic analysis of major heterochromatic elements especially Xh and Y) in *Drosophila melanogaster*, and the theory of "Heterochromatin". Chromosoma **10**, 535—588 (1959).
CROUSE, H. V.: The role of edysone in DNA-puff formation and DNA synthesis in the polytene chromosomes of *Sciara corprophila*. Proc. nat. Acad. Sci. **61**, 971—978 (1968).
— KEYL, G.: Extra replications in the "DNA-Puffs" of *Sciara corprophila*. Chromosoma (Berl.) **25**, 357—364 (1968).
DA CUNHA, A. B., PAVAN, C., MORGANTE, J. S., GARRIDO, M. C.: Studies on cytology and differentiation in Sciaridae. II. DNA redundancy in salivary gland cells of *Hybosciara fragilis*. Genetics Suppl. **61**, 335—349 (1969).
DANIELI, G. A., RODINO, E.: Larval moulting cycle and DNA synthesis in *Drosophila hydei* salivary glands. Nature (Lond.) **213**: 424—425 (1967).
DARROW, J. M., CLEVER, U.: Chromosome activity and cell function in polytenic cells. III. Growth and replication. Develop. Biol. 21, 331—348 (1970).
DERKSEN, J., BERENDES, H. D.: Polytene chromosome structure at the submicroscopic level. II. Length distribution of DNA molecules from polytene chromosomes of *Drosophila melanogaster* and *D. hydei*. Chromosoma **31**, 468—477 (1970).
DICKSON, E., BOYD, J. B., LAIRD, C. D.: Sequence diversity of polytene chromosome DNA from *Drosophila hydei*. J. molec. Biol. **61**, 615—627 (1971).
DION, A. S., HERBST, E. J.: Polyamine changes during development of *Drosophila melanogaster*. Ann. N. Y. Acad. Sci. **171**, 723—734 (1970).
ECKHARDT, R. A., GALL, J. G.: Satellite DNA associated with heterochromatin in Rhynchosciara. Chromosoma (Berl.) **32**, 407—427 (1971).
EDSTRÖM, J-E.: Chromosomal RNA and other nuclear RNA fractions. In: LOCKE, M., (Ed.): The Role of Chromosomes in Development, pp. 137—152, New York: Academic Press 1964.
FICQ, A., PAVAN, C., BRACHET, J.: Metabolic processes in chromosomes. Exp. Cell Res. **6**, 105—114 (1958).
FOX, D. P.: A non-doubling DNA series in somatic tissues of the locusts *Schistocerca gregaria* (Forskål) and *Locusta migratoria* (Linn.). Chromosoma (Berl.) **29**, 446—461 (1970).
— The replicative status of heterochromatic and euchromatic DNA in two somatic tissues of *Dermestes maculatus* (Dermestidae: Coleoptera). Chromosoma (Berl.) **33**, 183—195 (1971).
GABRUSEWYCZ-GARCIA, N.: Cytological and autoradiographic studies in *Sciara coprophila* salivary gland chromosomes. Chromosoma (Berl.) **15**, 312—344 (1964).

Gabrusewycz-Garcia, N., Kleinfeld, R. G.: A study of the nucleolar material in *Sciara coprophila*. J. Cell Biol. **29**, 347—359 (1966).

Gall, J. G., Cohen, E. H., Polan, M. L.: Repetitive DNA sequences in Drosophila chromosomes. Chromosoma (Berl.) **33**, 319—344 (1971).

— Pardue, M. L.: Formation and detection of RNA-DNA hybrid molecules in cytological preparations. Proc. nat. Acad. Sci. (Wash.) **63**, 378—383 (1969).

Gay, H., Das, C. C., Forward, K., Kaufmann, B. P.: DNA content of mitotically-active condensed chromosomes of *Drosophila melanogaster*. Chromosoma (Berl.) **32**, 213—223 (1970).

Gerbi, S. A.: Localization and characterization of the ribosomal RNA cistrons in *Sciara coprophila*. J. molec. Biol. **58**, 499—511 (1971).

Gorovsky, M. A., Woodard, J.: Histone content of chromosomal loci active and inactive in RNA synthesis. J. Cell Biol. **33**, 723—728 (1967).

Hadorn, E.: An accelerating effect of normal "Ring-glands" on pupariumformation in lethal larvae of *Drosophila melanogaster*. Proc. nat. Acad. Sci. (Wash.) **23**, 478—484 (1937).

— Ruch, F., Staub, M.: Zum DNS-Gehalt im Speicheldrüsenkernen mit (übergroßen Riesenchromosomen) von *Drosophila melanogaster*. Experientia (Basel) **20**, 566—567 (1964).

Hagele, K.: DNS-Replikationsmuster der Speicheldrüsen-Chromosomen von Chironomiden. Chromosoma (Berl.) **31**, 91—138 (1970).

Halfer, C., Tiepolo, L., Barigozzi, C., Fraccaro, M.: Timing of DNA replication of translocated Y chromosome sections in somatic cells of *Drosophila melanogaster*. Chromosoma (Berl.) **27**, 395—408 (1969).

Heitz, E.: Die somatische Heteropyknose bei *Drosophila melanogaster* und ihre genetische Bedeutung. Z. Zellforsch. mikr. Anat. **20**, 237—287 (1933).

Helmsing, P. J.: Protein synthesis of polytene nuclei *in vitro*. Biochim. biophys. Acta (Amst.) **224**, 579—587 (1970).

Henderson, A., Ritossa, F.: On the inheritance of rDNA of magnified *bobbed* loci in *D. melanogaster*. Genetics **66**, 463—473 (1970).

Henderson, S. A.: The salivary gland chromosomes of *Dasyneura crataegi* (Diotera: Cecidomyiidae). Chromosoma (Berl.) **23**, 38—58 (1967).

Hennig, W., Hennig, I., Stein, H.: Repeated sequences in the DNA of Drosophila and their localization in giant chromosomes. Chromosoma (Berl.) **32**, 31—63 (1970).

— Meer, B.: Reduced polyteny of ribosomal RNA cistrons in giant chromosomes of *Drosophila hydei*. Nature New Biol. **233**, 10—12 (1971).

Howard, E. F., Plaut, W.: Chromosomal DNA synthesis in *Drosophila melanogaster*. J. Cell Biol. **39**, 415—429 (1968).

Huberman, J. A., Riggs, A. D.: On the mechanism of DNA replication in mammalian chromosomes. J. molec. Biol. **32**, 327—341 (1968).

John, B., Lewis, K. R.: The Chromosome Cycle, pp. 60—62, Wien: Springer 1962.

John, H. A., Birnstiel, M. L., Jones, K. W.: RNA-DNA hybrids at the cytological level. Nature (Lond.) 223, 582—587 (1969).

Jones, K. W., Robertson, F.: Localisation of reiterated nucleotide sequences in *Drosophila* and mouse by *in situ* hybridisation of complementary RNA. Chromosoma (Berl.) **31**, 331—345 (1970).

Kaufmann, B. P.: Synthesis. In: "Genetics Today" (Proc. XI Int. Congr. Genet., The Hague, The Netherlands, Sept. 1963,) Vol. **2**, pp. 385—389. London: Pergamon Press 1965.

— Gay, H., Buchanan, J., Weingart, A., Maruyama, K., Akey, A.: Organization of cellular materials. Carnegie Inst. Wash. Yearbook **61**, 466—474 (1962).

Keyl, H.-G.: Verdopplung des DNS-Gehalts kleiner Chromosomenabschnitte als Faktor der Evolution. Naturwissenschaften **51**, 46—47 (1964).

— Lokale DNS-Replikationen in Riesenchromosomen. In: Sitte, P., (Ed.): Probleme der biologischen Reduplikation, S. 55—69. Berlin-Heidelberg-NewYork: Springer 1966.

— Hägele, K.: Heterochromatin-Proliferation an den Speicheldrüsen-Chromosomen von *Chironomus melanotus*. Chromosoma (Berl.) **19**, 223—230 (1966).

— Pelling, C.: Differentielle DNS-Replikation in den Speicheldrüsen-Chromosomen von *Chironomus thummi*. Chromosoma (Berl.) **14**, 347—359 (1963).

KRISHNAKUMARAN, A., BERRY, S. J., OBERLANDER, H., SCHEIDERMAN, H. A.: Nucleic acid synthesis during insect development—II. Control of DNA synthesis in the Cecropia silkworm and other saturniid moths. J. Insect Physiol. **13**, 1—57 (1967).

LAIRD, C. D.: Chromatid structure: Relationship between DNA content and nucleotide sequence diversity. Chromosoma (Berl.) **32**, 378—406 (1971).

LAKHOTIA, S. C., MUKHERJEE, A. S.: Chromosomal basis of dosage compensation in *Drosophila*. III. Early completion of replication by the polytene X-chromosome in male: further evidence and its implications. J. Cell Biol. **47**, 18—33 (1970).

LARA, F. J. S., HOLLANDER, F. M.: Changes in RNA metablism during the development of *Rhynchosciara angelae*. Nat. Cancer Inst. Monogr. **27**, 235—242 (1967).

LAUFER, H., NAKASE, Y., VANDERBERG, J.: Developmental studies of the dipteran salivary gland. I. The effects of actinomycin D on larval development, enzyme activity, and chromosomal differentiation in *Chironomus thummi*. Develop. Biol. **9**, 367—384 (1964).

LIMA-DE-FARIA, A.: DNA replication and gene amplification in heterochromatin. In: Lima-de-Faria, A., (Ed.): Handbook of Molecular Cytology, pp. 317—325. Amsterdam-London: North-Holland 1969.

MAHOWALD, A. P.: Polar granules of Drosophila II. Ultrastructural changes during early embryogenesis. J. exp. Zool. **167**, 237—261 (1968).

MATTINGLY, E., PARKER, C.: Nucleic acid synthesis during larval development of *Rhynchosciara*. J. Insect Physiol. **14**, 1077—1083 (1968).

MENEGHINI, R., ARMELIN, H. A., BALSAMO, J., LARA, F. J. S.: Indication of gene amplification in *Rhynchosciara* by RNA-DNA hybridization. J. Cell Biol. **49**, 913—916 (1971).

MULDER, M. P., VAN DIUJN, P., GLOOR, J. H.: The replicative organization of DNA in polytene chromosomes of *Drosophila hydei*. Genetica **39**, 385—428 (1968).

MULLER, H. J.: The development of the gene theory. In: Dunn, L. C., (Ed.): Genetics in the 20th Century, pp. 77—99. New York: Macmillan 1951.

— CARLSON, E., SCHALET, A.: Mutation by alteration of the already existing gene. Genetics **46**, 213—226 (1961).

NAGL, W.: Banded polytene chromosomes in the legume *Phaseolus vulgaris*. Nature (Lond.) **221**, 70—71 (1969).

NASH, D., BELL, J.: Larval age and the pattern of DNA synthesis in polytene chromosomes. Can. J. Genet. Cytol. **10**, 82—90 (1968).

— PLAUT, W.: On the presence of DNA in larval salivary gland nucleoli in *Drosophila melanogaster*. J. Cell Biol. **27**, 682—686 (1965).

— PARDUE, M. L., GERBI, S. A., ECKHARDT, R. A., GALL, J. G.: Cytological localization of DNA complementary to ribosomal RNA in polytene chromosomes of Diptera Chromosoma (Berl.) **29**, 268—290 (1970).

OLVERA, R. O.: The nucleolar DNA of three species of Drosophila in the *hydei* complex. Genetics Suppl. 1 **61**, 245—249 (1969).

PAVAN, C.: Chromosomal differentiation. Nat. Cancer Inst. Monogr. **18**, 309—323 (1965a).

— Nucleic acid metabolism in polytene chromosomes and the problem of differentiation. Brookhaven Symp. Biol. **18**, 222—241, (1965b).

— DA CUNHA, A. B.: Chromosome activities in normal and in infected cells of *Sciaridae*. Nucleus (Calcutta) Suppl. **12**, 183—196 (1968).

— — Chromosomal activities in Rhynchosciara and other sciaridae. Ann. Rev. Genet. **3**, 425—450 (1969a).

— — Gene amplification in ontogeny and phylogeny of animals. Genetics Suppl. **61**, 289—304 (1969b).

PELLING, C.: A replicative and synthetic chromosomal unit — the modern concept of the chromomere. Proc. roy. Soc. B **164**, 279—289 (1966).

— Synthesis of nucleic acids in giant chromosomes. In: BUTLER, J. A. V., NOBLE, D. (Eds.): Progress in Biophysics and Molecular Biology, pp. 239—270. London: Pergamon Press 1969.

PENNER, P. E., COHEN, L. H., LOEB, L. A.: RNA-dependent DNA polymerase: presence in normal human cells. Biochem. Biophys. Res. Commun. **42**, 1228—1234 (1971).

— — — RNA-dependent DNA polymerase in human lymphocytes during gene activation by phytohemagglutinin. Nature New Biol. **232**, 58—61 (1971).

Pettit, B. J., Rasch, R. W.: Tritiated-histidine incorporation into Drosophila salivary chromosomes. J. Cell Physiol. **68**, 325—334 (1966).
— — Rasch, E. M.: DNA synthesis in the giant salivary chromosomes of *Drosophila virilis* prior to pupation. J. Cell Physiol. **69**, 273—280 (1967).
Plaut, W.: On the replicative organization of DNA in the polytene chromosome of *Drosophila melanogaster*. J. molec. Biol. **7**, 632—635 (1963).
Prescott, D. M.: The structure and replication of eukaryotic chromosomes. In Adv. Cell Biol. **1**, 57—117 (1970).
Rae, P. M. M.: Chromosomal distribution of rapidly reannealing DNA in *Drosophila melanogaster*. Proc. nat. Acad. Sci. (Wash.) **67**, 1018—1025 (1970).
Rasch, E. M., Barr, H. J., Rasch, R. W.: The DNA content of sperm of *Drosophila melanogaster*. Chromosoma (Berl.) **33**, 1—18 (1971).
Ristow, H., Arends, S.: A system *in vitro* for the synthesis of RNA and protein by isolated salivary glands and by nuclei from Chironomus larvae. Biochim. biophys. Acta (Amst.) **157**, 178—186 (1968).
Ritossa, F. M.: Behaviour of RNA and DNA synthesis at the puff level in salivary gland chromosomes of Drosophila. Exp. Cell Res. **36**, 151—523 (1964).
Roberts, P. A., Kimball, R. F., Pavan, C.: Response of Rhynchosciara chromosomes to microsporidian infection. Increased polyteny and generalized puffing. Exp. Cell Res. **47**, 408—422 (1967).
Rodman, T. C.: DNA replication in salivary gland nuclei of *Drosophila melanogaster* at successive larval and prepupal stages. Genetics **55**, 376—386 (1967a).
— Control of polytenic replication in Dipteran larvae. I. Increased number of cycles in a mutant strain of *Drosophila melanogaster*. J. Cell. Physiol. **70**, 179—186 (1967b).
— Relationship of developmental stage to initiation of replication in polytene nuclei. Chromosoma (Berl.) **23**, 271—287 (1968).
Rudkin, G. T.: Cytochemistry in the ultraviolet. Microchem. J. Symp. Ser. **1**, 261—276 (1961).
— The proteins of polytene chromosomes. In: Bonner, A., Ts'o, A. (Eds.): The Nucleohistones, pp. 184—192. San Francisco: Holden Day 1964.
— The structure and function of heterochromatin. Genetics Today. (Proc. of XI Int. Congr. of Genetics, The Hague, The Netherlands, 1963), Vol. **2**, pp. 359—374. London: Pergamon Press 1965a.
— The relative mutabilities of DNA in regions of the X chromosoma of *Drosophila melanogaster*. Genetics **52**, 665—681 (1965).
— Non replicating DNA in Drosophila. Genetics, Suppl. **61**, 227—238 (1969).
— Corlette, S. L.: Disproportionate synthesis of DNA in a polytene chromosome region. Proc. nat. Acad. Sci. (Wash.) **43**, 964—968 (1957).
— Woods, P. S.: Incorporation of H^3 cytidine and H^3 thymidine into giant chromosomes of Drosophila during puff formation. Proc. nat. Acad. Sci. (Wash.) **45**, 997—1003 (1959).
Sauaia, H., Laicine, E. M., Alves, M. A. R.: Hydroxyurea-induced inhibition of DNA puff development in the salivary gland chromosomes of *Bradysia hygida*. Chromosoma (Berl.) **34**, 129—151 (1971).
Schultz, J.: The nature of heterochromatin. Cold Spring Harb. Symp. quant. Biol. **12**, 179—191 (1947).
— Genes, differentiation, and animal development. Brookhaven Symp. Biol. **18**, 116—147 (1965).
— Rudkin, G.: Direct measurement of deoxyribonucleic acid content of genetic loci in Drosophila. Science **132**, 1499—1500 (1960).
Skoog, L., Nordenskjöld, B.: Effects of hydroxyurea and 1-β-arabinofuranosyl cytosine on deoxyribonucleotide pools in mouse embryo cells. Europ. J. Biochem. **19**, 81 (1971).
Sonnenblick, B. P.: The early embryology of *Drosophila melanogaster*. In: Demerec, M. (Ed.): Biology of Drosophila, pp. 62—167. New York: John Wiley & Sons 1950.
Staub, M.: Veränderungen im Puffmuster und das Wachstum der Riesenchromosomen in Speicheldrüsen von *Drosophila melanogaster* aus spatlarvalen und embryonalen Spendern nach Kultur *in vivo*. Chromosoma (Berl.) **26**, 76—104 (1969).

Swift, H.: Nucleic acids and cell morphology in Dipteran salivary glands. In: Allen, J. M. (Ed.): The Molecular Control of Cellular Activity, pp. 73—125. New York: McGraw-Hill 1962.

— The histones of polytene chromosomes. In: Bonner, J., Ts'o, P. (Eds.): The Nucleohistones. pp. 169—183. San Francisco: Holden-Day 1964.

— Molecular morphology of the chromosome. In Vitro 1, 26—49 (1965).

— Nulear physiology and differentiation: a general summary. Genetics Suppl. 61, 439—461 (1969).

Swindlehurst, M., Berry, S. J., Firshein, W.: The biosynthesis of DNA by insects. III. Ribonucleotide reductase activity and the effect of hydroxyurea during the adult development of the cecropia silkworm. Biochim. biophys. Acta (Amst.) 228, 313—323 (1971).

Tartof, K.: Increasing the multiplicity of ribosomal RNA genes in Drosophila melanogaster. Science 171, 294—297 (1971).

Taylor, J. H.: Rates of chain growth and units of replication in DNA of mammalian chromosomes. J. molec. Biol. 31, 579—594 (1968).

— The structure and duplication of chromosomes. In: Caspari, E. W. (Ed.): Genetic Organization; a Comprehensive Treatise, pp. 163—221. New York-London: Academic Press 1969.

Temin, H. M., Mitzutani, S.: RNA-dependent DNA polymerase in virions of Rous Sarcoma Virus. Nature (Lond.) 226, 1211—1213 (1970).

Von Borstel, R. C., Prescott, D. M., Bollum, F. J.: Incorporation of nucleotides into nuclei of fixed cells by DNA polymerase. J. Cell Biol. 29, 21—28 (1966).

— Miller, O. L. jr., Bollum, F. J.: Probing the structure of chromosomes with DNA polymerase and terminal transferase. Genetics Suppl. 61, 401—408 (1969).

Wagner, R. P., Ed: (Proc. Int. Symp., Nuclear Physiology and Differentiation, Belo Horizonte, Minas Gerais, Brazil, December 1968). Genetics Suppl. 61, (1969).

Walker, P. M. B., Flamm, W. G., McLaren, A.: Highly repetitive DNA in rodents. In: Lima-de-Faria, A. (Ed.): Handbook of Molecular Cytology, pp. 52—66. Amsterdam-London: North Holland 1969.

Wanka, F., Moors, J., Krijzer, F. N. C. M.: Dissociation of nuclear DNA replication from concomitant protein synthesis in synchronous cultures of chlorella. Biochim. biophys. Acta (Amst.) (in press) (1972).

Welch, R. M.: A developmental analysis of the lethal mutant L(2)GL of Drosophila melanogaster based on cytophotometric determination of nuclear desoxyribonucleic acid (DNA) content. Genetics 42, 544—559 (1957).

—. Resch, K.: A cytochemical analysis of deoxribonucleic acid (DNA) and protein in salivary gland and gut of the lethal mutant lgl of D. melanogaster. Univ. Texas Pub. No. 6818, Studies in Genetics, No. 4, 49—70 (1968).

Whitten, J. M.: Differential deoxyribonucleic acid replication in the giant foot-pad cells of Sarcophaga bullata. Nature (Lond.) 208, 1019—1021 (1965).

Wolf, B. E.: Temperaturabhängige Allozyklie des polytänen X-Chromosoms in den Kernen der Somazellen von Phryne cincta. Chromosoma (Berl.) 8, 396—435 (1957).

— Y-Chromosom und überzählige Chromosomen in den polytänen Somakernen von Phryne cincta fabr. (Diptera, Nematocera). Verh. dtsch. Zool. Ges. Saarbrücken 1961, 110—123.

— Structure and function of alpha- and beta-heterochromatin-results on Phryne cincta. Nucleus (Calcutta) Suppl. 12, 145—160 (1968).

— Adaptiver chromosomaler Polymorphismus und flexible Kontrolle der Rekombination bei Phryne cincta (Diptera, Nematocera). Zool. Beitr. 14, 125—153 (1968).

Wolstenholme, D. R., Dawid, I. B., Ristow, H.: An electron microscope study of DNA molecules from Chironomus tentans and Chironomus thummi. Genetics 60, 759—770 (1968).

Young, C. W., Schochetman, G.: Hydroxyurea-induced inhibition of deoxyribonucleotide syntheses: studies in intact cells. Cancer Res. 27 (Part I), 526—534 (1967).

Transcription in Giant Chromosomal Puffs

CLAUS PELLING

Max-Planck-Institut für Biologie, Tübingen

I. Puffing

Since 1952, specific reversible changes in chromosomal structure have been correlated with the activity of genes (BEERMANN, 1952). This phenomenon, called puffing, indicated that the cell is able to control metabolic functions by turning genes on and off. Considerably later the concept of the genetic code and its actualization by the process of transcription and translation was to enter common consciousness.

The significance of puffing became apparent from observations by which a relationship between the presence of a puff and a cellular phenotype could be established.

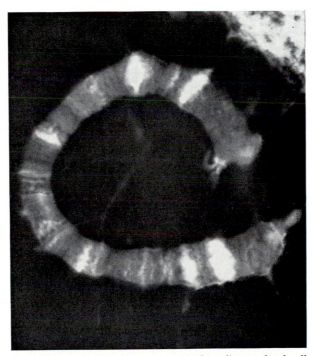

Fig. 1. Puffs of the ordinary type in chromosome I of a salivary gland cell of Chironomus tentans (Original). a RNA specific staining with toluidine blue, dark field

1) Thus different cell types which by definition have to have different cellular functions showed remarkable differences in their puff pattern.

2) In a given cell type temporal changes in cellular activities can be correlated with particular puffs. Such a case was realized when specific puffs were observed to appear during certain periods of Dipteran larval development.

3) Specific cellular products (e.g. salivary proteins) which may be clearly distinguished by morphological as well as by biochemical criteria have been demonstrated as being dependent on the presence of a particular puff (BEERMANN, 1961; GROSSBACH, 1969).

Puffs are structures to which the chromosomal synthesis of RNA is apparently restricted (PELLING, 1964). The specificity of this RNA is reflected in the base composition difference between RNAs taken from different puffs (EDSTRÖM and BEERMANN, 1962). In the salivary gland chromosomes of Chironomus tentans, with which

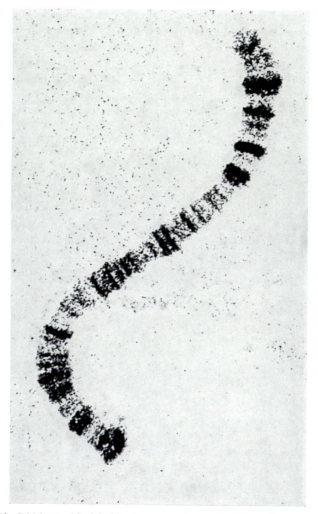

Fig. 1b. RNA specific label in the autoradiograph. (From PELLING, 1964)

this article will mainly deal, a distinction can be made between puffs of the ordinary type (Fig. 1) and Balbiani rings (Fig. 2). Typical puffs which constitute the majority of the active sites are comparatively small: In most cases the degree of puffing, i.e. the swelling of the structure and the concomittant disintegration of the chromosomal region remains moderate. Most Chironomus puffs preserve more or less the appearance of a chromosomal band. In such structures the intensity of uptake of radioactive RNA precursors is low in autoradiographs (Fig. 1b). Contrary to this, the Balbiani rings are extraordinarily enlarged and specialized puffs, sometimes assuming 3 times the width of the chromosome. Such a structure is able to carry out up to 15% of the total cellular RNA synthesis (Fig. 2b). The Balbiani rings are specific for the salivary gland cells; they apparently code for polypeptides of the salivary secretory protein (GROSSBACH, 1969).

Analyses of puff patterns provide information about the extent of gene activation in a differentiated cell. They also permit a subdivision of puffs into different categories. In order to illustrate this some quantitative details may be given. In the salivary gland chromosomes of the last larval instar of Chironomus tentans, in which about 2000 chromosomal bands have been analysed (BEERMANN, 1952) only 300 puffs can be

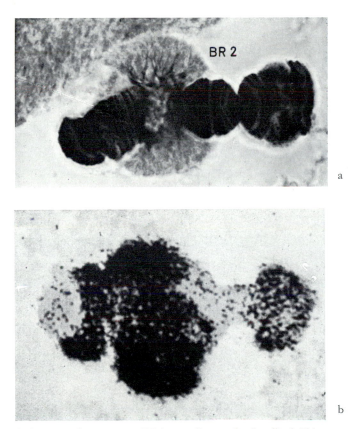

Fig. 2. Balbiani rings on chromosome IV in a salivary gland cell of Chironomus tentans (From PELLING, 1964). a Carmine-orcein squash preparation, BR 2 indicated. Magnification ca. 900 x. b RNA specific label in the autoradiograph, magnification ca. 800 x

discovered (PELLING, 1964). In these terms, then, 15% of the genome has to be regarded as active in a given tissue. Most of these structures are, of course, not tissue specific. The bulk of puffs present in one cell type can also be distinguished in others; in fact, puff pattern analyses performed in Drosophila hydei revealed that only 5% of the puffs are tissue specific (BERENDES, 1966). Obviously, the majority of the genetic information transcribed is needed for the maintainance of general metabolic functions in every cell, but from the numbers given above it appears that with respect to the total genetic information in the cell the fraction involved in these functions is relatively small. If each cell type has a tissue specific set of puffs amounting to 0.75% of the total number of bands, an upper number of cell types can be calculated for which the untranscribed rest of 85% of the bands can be used up. Thus about 120 "undifferentiated" cell types could be supplemented with their tissue specific information. This number does not seem to be unrealistic an estimate. It is left to the reader to agree or not that a Chironomus larva and a Chironomus fly can be constructed out of 120 different tissues.

Although, as already mentioned, it has been possible to relate the synthesis of a specific protein to a specific puff (GROSSBACH, 1969), there is no information available as to whether a puff contains more genetic information than is required for the synthesis of a single polypeptide. The small number of puffs found in the chromosome complement of a cell and also the extremely small number of tissue-specific puffs rather favour the concept that there is more than one function coded for per puff.

II. DNA Involved in Transcription

Some basic information required for our discussion concerns the amount of DNA which is involved in the process of transcription. To approach the problem first in cytological terms we can ask how much DNA is involved in the formation of a puff. During the course of puffing, chromosomal regions loose their structural definition, but with appropriate microscopical preparations the amount of DNA which "disappears" in the chromosomal area can be determined, thus providing an upper limit to the amount of active DNA in the region. However, the determination of the border of a puff presents a problem. For example, there are cases known in which only subsections of chromosomal bands puff (KEYL, 1965; PELLING, 1964). On the other hand, Balbiani ring 2 on chromosome IV of Chironomus may easily disintegrate 10 adjacent bands and interbands on either side of the chromosome; in other words such a structure can physically include the DNA of almost 1/6 of all bands of the particular chromosome. In the latter case there is fortunately some evidence that the process of chromosome disintegration does not automatically lead to a gene activation of all the DNA involved: Within a Balbiani ring area bundles of chromatids (oligonemic fibrils) can often be distinguished in which the chromomeric pattern is still intact and resembles the pattern of the unpuffed chromosome region. Local specificities can be determined if the puff is traced back to its initial stages, and it turns out that even the largest ones, as for instance nucleoli and Balbiani rings of Chironomus originate from a single band (BEERMANN, 1952). One band or even a portion of a band starts puffing and this gradually expands into the area of adjacent bands. No case has been observed in which a puff develops by a simultaneous disintegration of several bands. Dogmatically, the implication of this observation is expressed by the statement: The transcribing section proper of a puff, even during its largest extension, always remains restricted to the DNA of the band from which it originates.

It is not difficult to calculate the DNA amount of an average chromomere which is the haploid equivalent of the polyploid giant chromosomal band. In Chironomus, for example, the number of bands is 2000 and the DNA of the haploid genome amounts to 2×10^{-13} g, according to EDSTRÖM (1964). The value reached is a DNA amount of 10^{-16} g per haploid chromomere corresponding to a molecular weight of 60 million Daltons.

III. The Lampbrush Chromosome Loop as a Model for Transcription

Our present concept of the transcription process in higher organisms was very much stimulated by the excellent electron microscopic work of MILLER and his coworkers. He was able to specifically isolate and visualize the transcriptory apparatus of specific chromosomal regions in salamander oocytes. By his methods the spatial relationships between DNA, RNA-polymerase and the transcription product could be studied in situ. Fig. 3 illustrates such a situation in a lampbrush chromosome loop.

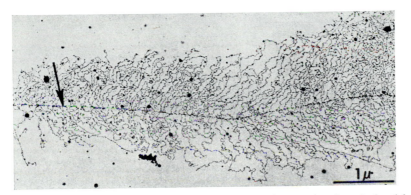

Fig. 3. Electron micrograph of a portion of a lampbrush chromosome loop isolated from a Triturus viridescens oocyte showing the DNA loop axis (arrow) with its gradient of attached RNP fibrils. (Courtesy of Dr. O. L. MILLER, JR., Oak Ridge)

Ribonucleoprotein fibrils are attached to a central DNA axis with their length increasing in one direction. In the electron microscope the structure repeats on a molecular level the same asymmetry ("polarisation") which also in the light microscope is an essential feature of the lampbrush loop (CALLAN and LLOYD, 1960). Particularly striking in this picture is the length of the RNA molecules produced. Although they appear shorter in comparison to the length of the DNA section from which they must have originated thus having undergone already a process of contraction (MILLER et al., 1970), lengthes of tens of microns have been reported. These observations certainly suggest that the DNA in a lampbrush loop constitutes extremely few and probably only one unit of transcription.

A second feature of the system is no less remarkable: The tremendous amount of RNA visible on the loop. It even becomes more striking when described in quantitative terms. The amount of RNA is directly correlated with the density of polymerases along the DNA axis, which in the illustration presented is about 40 polymerases per micron of DNA. On an average loop of 30 micron length this would multiply up to 1200 polymerases per loop, each of which would be making an RNA molecule. Provided that the loop contains no more than one unit of transcription, the mean length of an RNA molecule equals half of the loop length. Thus, the total amount of RNA attached to the loop would be 1200 molecules of an average length of 15 microns. In other words, the RNA of a loop exceeds the amount of DNA by a factor of 300.

As a general quantitative expression which also takes into account the possibility that there can be more than one DNA-segment active in transcription within one loop, we can write:

$$g\,RNA = \frac{l^2_{DNA}}{2\,n_T} \times \varrho_P \times \varrho_{RNA}$$

The RNA amount depends directly on the half of the square of the DNA length (l_{DNA}) times the number of polymerases per unit element (ϱ_P) times the weight of the RNA per unit element (ϱ_{RNA}). It depends reciprocally on the number of transcriptory elements (n_T). Regarding the number of transcriptory elements in a loop, this relationship holds only if the size distribution of the elements is not very uneven.

Fig. 4 compares a loop with one transcriptory element with another one, in which the same length of DNA is subdivided into three transcriptory elements identical in length. If all other parameters are kept constant, subdivision of the DNA into several transcribing segments leads to a decrease of RNA, in the example given by a factor of 3.

If the same considerations are applied to giant chromosome puffs some differences between them and lampbrush loops become apparent, in particular with respect to the RNA concentration at an active site: It is a common experience that the RNA in giant chromosomal puffs is rather difficult to distinguish by cytochemical techniques and far more difficult to determine. This would not be expected if an hundredfold or more excess of RNA over DNA were a common situation in a giant chromosome puff as it is in a lampbrush loop. There are no data bearing directly on the amount of RNA in individual puffs except for the Balbiani rings (see below), but it is possible to give an estimation based on the earlier microelectrophoretic work of Edström and Beermann (1962): Apparently a long chromosome like chromosome I in a salivary gland cell of Chironomus tentans contains as much RNA as one single large Balbiani ring on chromosome IV (20 μμg). The RNA in chromosome I, however, is distributed over about 70 puffs. The RNA amount of an average puff is therefore on the order of 0.3 μμg. Taking into account that the giant chromosome of Chironomus is a multiple structure containing thousands of identical chromatids (in most nuclei 4000 or 8000) quite different RNA to DNA ratios are obtained in comparison to the lampbrush loop: In a Balbiani ring there exists only 33 times more RNA than DNA. In a puff of average size, however, the DNA even exceeds the RNA by a factor of 2. Provided that the whole chromomeric DNA of a puff is involved in transcription and that n_T equals 1 then it follows that an ordinary puff operates with only two polymerases per strand. Even the gigantic Balbiani ring which is most

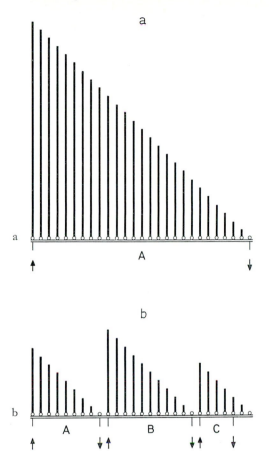

Fig. 4. Model of transcription in a lampbrush loop. a Entire DNA constitutes one unit of transcription. b DNA is subdivided into 3 transcriptory elements of equal length. = DNA, — RNA, ○ Polymerase, ↓ Initiation site, ↑ Termination site

certainly the largest puff structure seen in any Dipteran cell contains about 120 polymerases per strand. The density of the polymerase differs between the lampbrush system and the giant chromosome system by more than two orders of magnitude (ordinary puffs), at least, however, by one order of magnitude (Balbiani ring).

Perhaps it may be emphasized that the situation at the giant chromosomal puff should not be interpreted as exhibiting an abnormally low polymerase concentration. Rather it may be true that the lampbrush chromosomes are extraordinarily densely occupied with polymerases. They could hardly increase their polymerase concentration further. As MILLER points out in a consideration of the distribution of polymerase molecules on nucleolar genes in amphibia, a separation of polymerase molecules by a distance of 100—200 Angströms is not very far from physical saturation (MILLER and BEATTY, 1969).

The differences in polymerase density which have been pointed out above can be put into a context with other already known characteristics by which the two systems can be distinguished from each other. There is, first, the difference in the proportion of active chromosomal structures (puffs or loops). The lampbrush chromosome is a chromosome type in which apparently all chromomeres are activated along the whole length of the chromosome. In the giant chromosomes, however, only a small and selected proportion of sites is allowed to be active. Further differences exist in terms of the variability of the active structures. Loop formation in lampbrush chromosomes is a long term process involving months of growth and largely lacking individual differences between different loops. The essential character of the puffing phenomenon, however, is the rapidity and individuality of the response of genetic sites to external and internal stimuli. In the case of ecdyson induction, for example, the puff forming mechanism operates within a few minutes and with a striking specificity (CLEVER and KARLSON, 1960).

Saturation of polymerase density, uniformity of the gene activation process and the lack of specificity and speed of genic responses clearly emphasize how inflexible the transcription system in the oocyte nuclei is in terms of gene regulation at the chromosome level. It appears to lack those functions on which the high regulatory capacity of the polytene chromosomes are based. One would even dare to say that it appears as though the lampbrush chromosomes are bound to produce RNA largely according to quantity, giant chromosomes, however, according to quality.

It may finally be worthwhile to point out an important implication of our discussion. It is to draw attention to the molecular basis of the puffing process itself. So far, our comparisons concerned differences in RNA concentration between lampbrush chromosome loops and puffs. The difference in the amount of RNA of the same chromosomal band, when it is at different stages of puffing, can also be explained in terms of the difference in polymerase concentration. It has well been documented that a puff with increasing size gains additional RNA. Unless we change the number of transcriptory elements or the length of the DNA involved in transcription the puff can gain more RNA only by increasing its polymerase density. Unless we have to learn that the products of a puff become different in the course of puff growth we may regard the size of a puff as a measure for a particular polymerase density. Essentially, puffing may be the process of putting increasing numbers of polymerases unto the DNA strand. The disintegration of the chromomeric structure which usually, and presumably correctly, has been described as a gradual uncoiling of the chromomeric DNA may in sterical terms be understood as providing increasing space for an increasing number of polymerases together with the RNA molecules attached. Also the gradual decrease of the distance between neighbouring polymerases may impose increasing rigidity on the DNA strand and may thus cause its inability of maintaining a complex tertiary structure like the chromomeric structure.

IV. Length of the RNA Molecules Transcribed in a Puff

Our considerations have so far been based on two main assumptions, the first being that according to the electron microscopic evidence the transcribing system of lampbrush loops and giant chromosome puffs consists of a single continuously transcribing segment, and the second being that the whole DNA of the structure involved participates in the transcription.

What here is termed whole DNA of the structure, however, may not be equivalent in both systems: In the lampbrush chromosome this is the DNA which makes up the central axis of the loop, which in turn is only a minor portion of the total DNA of the particular chromomere out of which the loop has been unravelled (GALL, 1963). In the case of the puff, however, the actual portion of the chromomere which has undergone a comparable process of unwinding cannot be determined, the only value which therefore can be referred to is the entire chromomeric DNA.

At this stage it has to be admitted, that no evidence has been presented that both above assumptions hold in the case of the giant chromosome system and it is entirely clear that the above conclusions about the polymerase concentration at the active sites remain imaginary as long as the fundamental similarity between loops and puffs cannot be substantiated. The puff as a system, then, could reach any degree of complexity: If we allow the number of transcriptory elements to be increased or the length of DNA involved in RNA-synthesis to be decreased, then the number of polymerases on the template has to increase in order to account for the loss of RNA from the relevant structure (see p. 92). If however, the polymerase concentration were to remain unchanged then the difference in RNA amount could be accounted for in a different way. We could envisage the existence of a pool of finished molecules within a puff. These molecules may not have had the opportunity to leave the puff area immediately after their completion. Apparently the nucleolus is a structure operating under these conditions. The precursor molecules of ribosomal RNA seem to remain in the nucleolus for a while, obviously for reasons of processing the RNA molecules. Also more complex puff models can be suggested (see below). But before proceeding in this discussion further, it is to be asked how long the puff RNA molecules are. This information is essential in order to interpret the transcriptory property of an active site.

We already know from earlier analyses of nuclear heterogeneous RNA (SCHERRER et al., 1966) that the average molecular weights are fairly high, in the order of several million. If this figure would be in the range of 30 million, the chromomeric DNA would be likely to constitute a contineously transcribing segment, provided that the duck erythroblast system were largely comparable to the Chironomus system discussed here. The molecular weights, however, are too small, approximately by a factor of 3 to 5.

In the past few years data on the size distribution of RNA molecules from individual giant chromosomes and single active loci have also been obtained (DANEHOLT et al., 1970; PELLING, 1970).

For example, Fig. 5 illustrates a radioactivity profile of RNA, isolated from Balbiani ring II after a very short pulse of tritiated uridine and separated on an acrylamide gel. Fig. 6 demonstrates another profile from the same structure after a long term pulse. Separation has been carried out on an agarose gel. The main feature of these RNA separations is the complexity of the RNA. According to EDSTRÖM and DANEHOLT (1967), the sedimentation values range between 10 s and 90 s.

The most straightforward interpretation of such pictures regards the multitude of the RNA types as a reflection of the complexity of the transcriptory elements within a single puff region. Apart from the heterogeneity of the patterns, however, there is another characteristic of the profiles to be accounted for: These patterns obtained from Balbiani rings were not constant in themselves. Sometimes the radioactivity

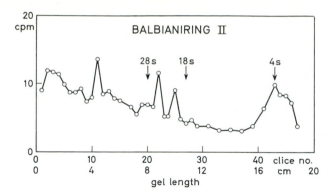

Fig. 5. Polyacrylamide gel separation of radioactive RNA from Balbiani ring 2, labelled for 2.5 min, polyacrylamide concentration 2.5%, separation time 6 h, location of reference peaks (28s and 18s) was determined in a parallel run, radioactivity was measured in the gas phase (From PELLING, 1970)

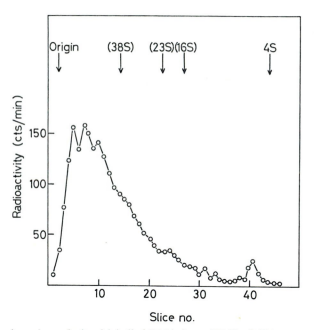

Fig. 6. Electrophoretic analysis of labelled RNA from BR II of Chironomus tentans. Salivary glands were incubated for 45 min in larval hemolymph, provided with tritiated cytidine and uridine. Fifty BR II were isolated, RNA extracted and analysed by electrophoresis in 2% agarose. (Courtesy of Dr. B. DANEHOLT, Stockholm)

focussed rather on the right side of the spectrum where the small molecules are expected to be localized, sometimes it shifted more to the left.

The pattern of Fig. 6 constitutes an example of the latter type. Attempts were made in order to explain this quasi-kinetic behaviour of the RNA spectra and several models were discussed (DANEHOLT et al., 1970; PELLING, 1970). One of them (Fig. 7)

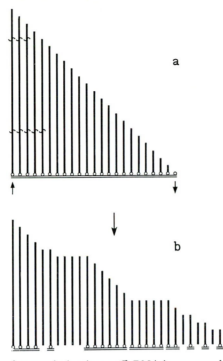

Fig. 7. Possible mode of transcription in a puff. RNA is processed at the chromosomal level
a) situation of the puff transcriptory unit b) array of molecules according to size, indicating
molecular heterogeneity of the system. ∼ cleavage point, for other symbols see legend of
Fig. 4. (Changed after PELLING, 1970)

postulated that a Balbiani ring is constructed of a single transcriptory unit as illustrated in Fig. 4a. But contrary to this simple version of the model, the ordered relationship between the length of DNA, the concentration of polymerases and the RNA
amount is altered by a mechanism which processes the RNA molecule on the RNA
template before its synthesis has come to an end. This modified model invokes the
action of a process which chooses the earliest possible moment to transport the RNA
away from the template.

   This model would place the mechanism of molecular processing, which has
to be postulated in order to account for the difference in molecular weight between
nuclear heterogeneous RNA and cytoplasmic "messenger RNA", to work — at least
at times — in the immediate neighborhood of the chromosomes. The model also
would give an explanation for the minuteness of the RNA amount present at a puff,
not necessarily by assuming a very low polymerase concentration on the DNA strand
as we have done before. The aforementioned heterogeneity and the "instability" of
the RNA patterns obtained may, however, also have originated from breakdown
— though a very slow, and judging from the reproducibility of peaks in the profiles
published, not unspecific breakdown — of a uniform RNA during handling and
sectioning of chromosomes and during the isolation procedure. This interpretation
has recently been suggested by DANEHOLT (personal communication), who obtained

RNA of particularly high molecular weight already approaching the value which is required if the identity between the whole chrommeric DNA and the transcribing portion of it is postulated. Such large molecules clearly reflect the tremendous length of the transcribing segment of the DNA within the Balbiani ring.

Still away from having clarified the chromosomal transcriptory apparatus in any details the data have, however, proceeded in substantiating the concept of the functional compartmentilisation of the chromosome of higher organisms presented earlier (BEERMANN, 1966; PELLING, 1966). The chromomeres, the linear array of which builds up the chromosome structure proper have been postulated to be units of replication and transcription. It appears now with respect to transcription that this postulate holds even in a more direct sense that one could have expected in molecular terms.

Interesting is one implication of this functional subdivision of the chromosome. If, at the chromosomal level, transcription occurs only in entities which are not smaller than chromomeres then the number of functional elements to be activated or blocked independently of each other cannot be higher in number than the number of chromosomal bands. If we imply that there are more coding sequences than there are chromosomal bands, restrictions are imposed on the system with respect to its regulatory capacity. This could be compensated for in two different ways. On the one hand, sequences which the cell simultaneously requires with each other could be arranged in clusters. On the other hand, if the cistrons were to be arranged randomly in the genome, the situation would, first, necessitate a simultaneous transcription of an — according to the ratio of coding sequences over chromomeres rapidly — increasing number of chromomeres. Secondly, a subsequent complex mechanism would have to be postulated to select the appropriate sequences out of the excess of the RNA produced.

Acknowledgements

I am indebted to Dr. B. DANEHOLT for providing me with Fig. 6: I am very grateful to Dr. O. L. MILLER, Jr. for his permission to use again the electron micrograph of the lampbrush loop and to Dr. P. RAE for invaluable help in preparing this manuscript. I also wish to thank Drs. K. GÖTZ and H. STEIN for stimulating discussion, Frl. H. BÜRGERMEISTER for most careful technical assistance and Herrn E. FREIBERG for executing the illustrations.

References

BEERMANN, W.: Chromosomenkonstanz und spezifische Modifikationen der Chromosomenstruktur in der Entwicklung und Organdifferenzierung von Chironomus tentans. Chromosome (Berl.) **5**, 139—198 (1952).
— Ein Balbianiring als Locus einer Speicheldrüsenmutation. Chromosoma (Berl.) **12**, 1—25 (1961).
BEERMANN, W.: Differentiation at the Level of the Chromosomes. In: Cell Differentiation and Morphogenesis pp. 24—54. North Holland Publ. Comp., Amsterdam 1966.
BERENDES, H. D.: Gene activities in the malpighian tubules of drosophila hydei at different developmental stages. J. exp. Zool. **162**, 209—217 (1966).
CALLAN, H. G., LLOYD, L.: Lampbrush Chromosomes of Crested Newts Triturus cristatus (Laurenti). Phil. Transact. Roy. Soc. (Lond.) B. **243**, 135—219 (1960).
CLEVER, U., KARLSON, P.: Induktion von Puff-Veränderungen in den Speicheldrüsenchromosomen von Chironomus tentans durch Ecdyson. Exp. Cell Res. **20**, 623—629 (1960).

DANEHOLT, B., EDSTRÖM, J.-E., EGYHAZI, E., LAMBERT, B., RINGBORG, U.: RNA synthesis in a Balbiani ring in chironomus tentans. Cold Spr. Harb. Symp. Quant. Biol. **35**, 513—519 (1970).

EDSTRÖM, J.-E.: Chromosomal RNA and other nuclear RNA fractions. Role of chromosomes in development 137—152. New York: Academic Press 1964.

— BEERMANN, W.: The base composition of nucleic acids in chromosomes, puffs, nucleoli, and cytoplasm of chironomus salivary gland cells. J. biophys. biochem. Cytol. **14**, 371—380 (1962).

— DANEHOLT, B.: Sedimentation properties of the newly synthesized RNA from isolated nuclear components of chironomus tentans salivary gland cells. J. molec. Biol. **28**, 331—343 (1967).

GALL, I. G.: Chromosomes and cytodifferentiation. Cytodifferential and macromolecular synthesis, pp. 119—143. New York: Academic Press 1963.

GROSSBACH, U.: Chromosomen-Aktivität und biochemische Zelldifferenzierung in den Speicheldrüsen von Camptochironomus. Chromosoma (Berl.) **28**, 136—187 (1969).

KEYL, H.-G.: Duplikationen von Untereinheiten der chromosomalen DNS während der Evolution von Chironomus thummi. Chromosoma (Berl.) **17**, 139—180 (1965).

MILLER, Jr., O. L., BEATTY, B. R.: Visualization of nucleolar genes. Science **164**, 955—957 (1969).

— — HAMKALO, B. A., THOMAS, Jr., C. A.: Electron microscopic visualization of transcription. Cold Spr. Harb. Symp. quant. Biol. **35**, 505—512 (1970).

PELLING, C.: Ribonukleinsäure-Synthese der Riesenchromosomen. Autoradiographische Untersuchungen an Chironomus tentans. Chromosoma (Berl.) **15**, 71—122 (1964).

— A replicative and synthetic chromosomal unit — the modern concept of the chromomere. Proc. roy. Soc. B **164**, 279—289 (1966).

— Puff-RNA in polytene chromosomes. Cold Spr. Harb. Symp. quant. Biol. **35**, 521—531 (1970).

SCHERRER, K., MARCAUD, L., ZAJDELA, F., LONDON, J. M., GROS, F.: Patterns of RNA metabolism in a differentiated cell: A rapidly labeled, unstable 60s RNA with messenger properties in duck erythroblasts. Proc. nat. Acad. Sci. (Wash.) **56**, 1571—1578 (1966).

Puffing Patterns in Drosophila Melanogaster and Related Species

MICHAEL ASHBURNER

Department of Genetics, University of Cambridge, Cambridge

I. Introduction

The major objective of this review is to summarize our knowledge of the behaviour of polytene chromosome puffs in *Drosophila melanogaster* and some of its more closely related species. It is hoped that this review will provide a sufficient background of information for further analysis of puffing in *D. melanogaster* for, as will become apparent later in this review, the wealth of our knowledge of the genetics of this species, and the relative ease with which it can be subjected to "genetic engineering", far outweigh its disadvantages as an object of cytological examination.

II. Puffing Patterns of *D. melanogaster*

The study of the puffs of *Drosophila* lagged behind their study in other groups such as *Chironomus* or *Rhyncosciara*. Although noticed by BRIDGES (1935) and studied by, for example, FREIRE-MAIA and FREIRE-MAIA (1953), RUDKIN (1955) and RUDKIN and WOODS (1959) the puffs of the salivary gland chromosomes of *Drosophila melanogaster* were almost totally neglected until the important paper of BECKER published in 1959. Working with the Berlin-normal "wild type" stock BECKER established that during the period of the L3[1] preceeding puparium formation and the subsequent prepupal period a large number of specific chromosome loci were involved in puff formation in a highly coordinate and regular manner. Studying in detail the left arm of chromosome three and the tip of the X chromosome he showed that until a time towards the end of the L3 few puffs were active. As the time of puparium formation approached the first changes occurred in this puffing pattern with the activation of puffs at 63F, 74EF and 75B on 3L and the regression of puffs active earlier in the instar at 68C on 3L and at 3C on the X chromosome. These dynamic changes in the pattern of puffed loci continued throughout the last part of the L3 and into the prepupal period. BECKER recognised that although these changes were part of a continuous process a number of Puff Stages (PS), each

---

1 The following abbreviations will be used: L1, L2, L3 first, second and third instar larvae respectively. PS Puff Stage. 2L, 2R, 3L and 3R the left and right arms of chromosomes 2 and 3 of *D. melanogaster*.

characterised by a unique pattern of puffs, could, for the experimentalist's convenience, be recognised. BECKER (1959) characterised 15 PS, puparium formation occurring at PS 10—11. Later ASHBURNER (1967a) extended this analysis to cover the complete prepupal period and described 21 PS of which the first 15 were almost identical to those of BECKER. Two further points were emphasised in BECKER's first paper on puffing: (a) that the time of puparium formation, and later of the prepupal/pupal moult, were periods of intense puffing activity, and (b) that several puffs were active at both of these developmental stages (although inactive in the intervening period) while other puffs were unique to one or other stage.

BECKER only analysed in detail the puffing patterns of the tip of the X chromosome and of 3L. This descriptive analysis was extended to cover all chromosome arms of *D. melanogaster* by ASHBURNER in two papers, the first dealing with the autosomes (1967a) and the second with the X chromosome (1969a). The general conclusions of BECKER's study were confirmed in these two papers which give a detailed description of the changes in puffing activity seen in the Oregon-R "wild type" stock from

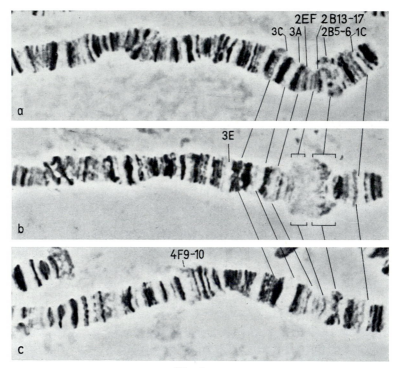

Fig. 1a—c

Fig. 1a—g. Puffing sequence of the distal end of *D. melanogaster* X chromosome. a PS1 (110 h larva, female), b PS8 (120 h larva, female), c PS10 (0 h prepupa, female), d PS15 (4 h prepupa, female), e PS18 (8 h prepupa, female), f PS20 (10 h prepupa, male), g PS21 (12 h prepupa, female). All Oregon. (From ASHBURNER, 1969a). Note on illustrations of the standard patterns (Figs. 1—16). Not all photographs are from accurately timed animals, the ages have been normalised to that of a strain with the mean age at puparium formation of 120 h (25° C). The photographs are from various strains, some wild type (Oregon, Barton, Canton-S and Pacific) and others carrying mutants the middle of the L3 to the time of complete salivary gland histolysis just after the prepupal/pupal moult.

Descriptions of these changes in puffing patterns make tedious reading and will not be repeated here. The reader is referred to the three papers mentioned above (BECKER, 1959; ASHBURNER, 1967a; 1969a) for their detail. Subsequent studies of the Oregon-R and other stocks of *D. melanogaster* necessitates some revision of the published patterns. For this reason revised tables of the PS for each chromosome arm are published in this paper and as much information as possible is given in the Figs. 1—16 which illustrate these patterns. A few explanatory notes follow.

Unlike the tables and figures of *D. melanogaster* puffing patterns previously published these are composite tables and do not refer to any particular stock. Although the basic

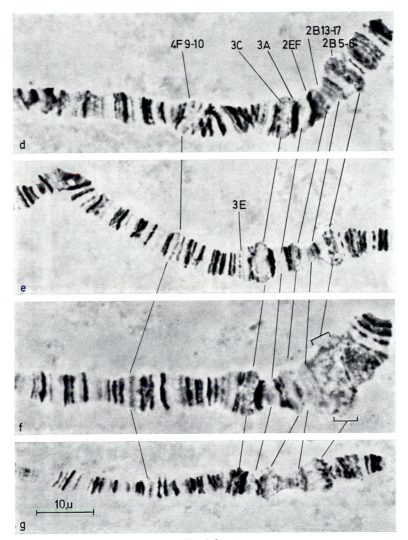

Fig. 1d—g

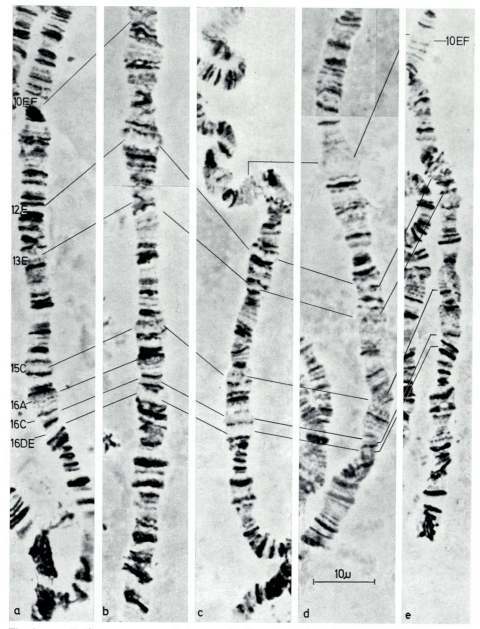

Fig. 2a—e. Puffing sequence of the proximal end of *D. melanogaster* X chromosome. a PS1 (110 h larva, female), b PS8 (120 h larva, female), c PS10 (0 h prepupa, female), d PS15 (4 h prepupa, female), e PS19 (10 h prepupa female). All Oregon. (From Ashburner, 1969a)

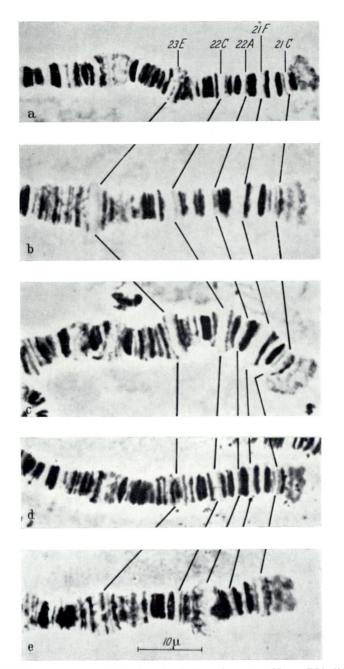

Fig. 3a—e. Puffing sequence of distal end of *D. melanogaster* 2L. a PS1 (110 h larva), b PS3 (115 h larva), c PS9 (120 h larva), d PS14 (4 h prepupa), e PS19 (10 h prepupa). All Oregon. (From ASHBURNER, 1967a)

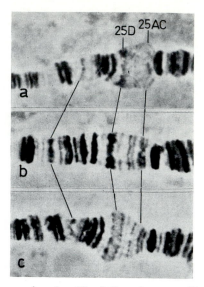

Fig. 4a—c. Puffing sequence of region 25 of *D. melanogaster* 2L. a PS1 (110 h larva, *vg6*). b PS2—3 (110 h larva, Oregon), c PS11—12 (¹/₂ h prepupa, *vg6* derived line)

data are from the Oregon-R stock information from several other stocks (especially the Canton-S, Pacific, and Barton "wild types") has been incorporated in order to give a more readily generalised picture of the puffing patterns of *D. melanogaster*. For instance in those cases where the activity of a particular locus (e.g. 22B4—5) in Oregon-R clearly differs from its activity in the majority of stocks examined the tables have been altered. The patterns defined by Tables 1—5 will be referred to as the "standard pattern" of *D. melanogaster*.

Table 1. Standard puff stages for X-chromosome

PS	1C	2B5—6	2B13—17	2EF	3A	3C	3E	4F9—10	7D	8D	8EF
1	—	+	—	—	—	++	+	—	—	(+)	—
2	—	++	—	—	—	+	(+)	—	—	(+)	—
3	—	++	(+)	—	—	—	(+)	—	—	—	—
4	—	++	+	+	—	—	—	—	—	—	—
5	—	++	++	+	—	—	—	—	—	—	—
6	(+)	++	++	+	—	—	—	—	—	—	—
7-8	(+)	++	++	+	+/—	—	—	—	—	—/+	—
9	(+)	++	++	+	+/—	—	(+)	(+)	—	+	—
10	(+)	(+)	+	+	+	—	(+)	(+)	—	+	—
11	(+)	—	+/—	++	++	—	(+)	+/++	+	+	+
12	(+)	+/—	+/—	++	+/(+)	—	—	+/++	+	(+)	+
13	—	+/—	+/—	+/—	+/—	—	—	+/++	+	(+)	(+)
14	—	+/—	—	+/—	—	(+)	(+)	+	+	—	—
15	—	+/(+)	—	+/—	—	+	(+)	+	—	—	—
16-17	—	—	—	—	—	+	(+)	—	—	—	—
18	—	—/+	—	—	(+)	++	+	—	—	—	—
19-20	—	++	(+)/+	+	(+)	++	(+)	(+)/+	—	—	—
21	(+)	++	—	++	—	—	—	(+)/+	—	—	—

Some of the major differences of this standard pattern from the Oregon-R data are, (a) incorporation of puffs absent from Oregon-R but generally distributed in other stocks (e.g. 22B4—5) and of a few extra puffs, all rather small, previously overlooked (e.g. 49B, 51D and 74C), (b) the fact that some puffs previously thought to be active only in prepupae are now recognised to show slight puffing in earlier PS also (e.g. 52C, 69A and 86E) and (c) in 1967 ASHBURNER stated that "despite a careful search no puffs have been seen on the small chromosome four". Subsequently a puff in 4 : 102 has been observed in 12 h. prepupae (PS 21) of several stocks (Fig. 16). Furthermore ROBERTS (1969) has recently reported the appearance of a puff in 4 : 101—102A when this chromosome was translocated to the tip of 3R following breakage in 4 : 101A. This puff is active in L3 and regresses at the time of puparium formation (ROBERTS, 1970).

When larvae are grown under standard conditions at 25° C the salivary gland chromosomes reach a stage amenable to cytogenetic analysis on the 4th day of larval development (70—80 h after oviposition). The puffing pattern of these young L3 is that described as PS1: relatively few sites are active, those at X: 3C, 2L: 25AC, 2R: 42A and 58DE, 3L: 68C and 3R: 85E and 90BC are the only prominent puffs. This pattern persists for a long time — indeed until about 10 h before puparium formation which occurs at approximately 120 h after oviposition (at 25° C). At this time the titre of moulting hormone (ecdysone) in the heamolymph increases and as a result of this the dramatic changes in puffing previously described are initiated. The hormonal control of puffing in *D. melanogaster* will be discussed in section V.

Peak periods of puffing activity, as measured by the number of loci puffed or as the mean size of those puffs active, occur at, or immediately before, puparium formation and in eight hour prepupae. It is of interest to note that each peak precedes the subsequent apolysis by just 4 h. The "prepupal moult" (from the pharate L3 to a pharate L4) occurs 4 h after puparium formation and the true pupal moult (from the pharate L4 to the pharate pupa) 12 h after puparium formation. The fact that many puffs are common to both of these periods, yet inactive during the intermould, while

(male/female when they differ)

9EF	10EF	11B	12E	13E1—2	14B	15C	16A	16C	16DE
—	+	++	—	+	—	+	—	—	—
—	+	++	—	(+)	—	+	+	—	—
—	+	+	—	(+)	—	+	+	—	—
—	(+)	+	—	—	—	+	+	—	—
—	(+)	+	—	—	—	+	—	—	(+)
—	—	+	+	—	—	+	—	—	+
—/+	+	+	+	—	(+)/—	+	—	—	+
(+)/+	++/+	+	+	—	(+)/—	+	—	+	+
+/++	++/+	++/+	++	+	(+)/—	+	—	+	+
+	++/+	++/+	++	+	+/(+)	+	—	+	(+)
+	++/+	++/+	—	+	+/(+)	+	—	+	—
+	++/+	++/—	—	+	(+)	+	—	—	—
+	++	+/—	—	+	—/(+)	(+)	—	—	—
+	++	+/—	—	+	—	(+)	—/(+)?	—	—
+	++	+/—	—	+	—	(+)/+	—	—	—
(+)	++/+	++/+	—/(+)	+	—	(+)/+	—	—	—
—	++/+	++/+	—	++	—	+/(+)	—	—	—
—	+/—	++/+	—	+	—	+/—	—	(+)	—

Table 2. The standard puff stages for 2L

PS	21C	21F	22A	22B4—5	22C	23E	25AC	25D	26B	27C	28A	29BC	29F	30A	30B	32C	33B	33E	34A	35B	36F	39BC
1	(+)						+									(+)			+			
2	(+)			(+)			+									(+)			+			
3				+												(+)			+			
4				+												(+)			+			
5		+		+	+											(+)	+	+	+			
6		+		+	+	+							+			(+)	+	+	+			+
7		+		+	+	+							+			(+)		+	+			+
8					+	+						+	+			(+)		+	+			+
9					+	+		+			+	+	+			(+)		+	+			+
10						+		+			+	+	+		(+)			+	+			+
11						+		+	(+)		+	+	+		(+)			(+)	+			+
12									(+)		+		+					(+)	+			
13											+		+					(+)	+			
14											+		+						+			
15								+		+	+		+						+	+		
16								+		+	+		+			+			+	+	+	
17										+	+			+		+			+	+	+	+
18										+	+			+		+			+	+	+	+
19	(+)		+		+	+		+	+	+	+	+		+		+	+	+	+	+	+	+
20	(+)	+	+			+		+	+	+	+	+				+	+	+	+			+
21	(+)					+		+	+	+	+	+					+	+	+			

Table 3. The standard puff stages for 2R

PS	42A	43E	44AB	46A	46F	47A	47BC	48B	49B	49E	50A	50CD	50F	51D	52A	52C	55B	55E	56D	57E	58BC	58DE	59F	60B
1	+	+	−	−	−	−	−	−	−	−	−	−	−	−	−	(+)	−	−	−	−	−	+	−	−
2	+	+	−	−	−	+	(++)	−	−	−	−	+	−	−	−	−	−	−	−	−	−	+	−	−
3	+	+	−	−	−	+	(++)	−	−	−	−	+	−	−	−	−	−	−	−	−	−	+	−	−
4	+	+	−	−	−	+	(++)	−	−	−	−	+	−	−	−	−	−	+	−	(++)	+	+	−	−
5	+	+	−	−	−	+	(++)	−	−	−	−	+	−	−	−	−	(++)	+	−	(++)	+	+	−	−
6	+	+	−	−	+	+	(++)	+	−	−	−	+	−	−	−	−	(++)	+	−	(++)	+	+	−	−
7	+	+	+	+	+	+	+	+	−	−	−	+	−	−	−	−	−	+	−	−	+	+	+	−
8	+	+	+	+	+	+	+	+	−	−	−	+	−	−	−	−	−	+	−	−	−	+	+	−
9	+	+	+	+	+	+	+	+	−	−	−	+	(++)	(++)	−	−	−	+	−	−	−	+	+	−
10	−	+	+	+	+	+	+	+	−	−	−	+	(++)	(++)	−	−	−	+	+	−	−	+	+	−
11	−	−	−	+	+	+	+	(+)	−	−	−	+	(++)	−	−	−	−	+	+	−	−	+	+	−
12	−	+	−	−	+	+	+	−	−	+	−	+	−	−	−	−	−	+	+	−	−	−	+	+
13	−	+	−	−	+	+	+	−	−	+	−	+	−	−	−	−	−	−	+	−	−	−	+	+
14	−	+	−	−	+	+	+	−	−	+	−	+	−	−	−	−	−	−	−	−	−	−	+	++
15	−	+	−	+	+	+	+	−	−	+	(++)	+	−	−	−	−	−	−	−	−	−	+	+	++
16	−	++	−	+	+	+	(++)	−	(++)	−	(++)	+	−	−	+	−	−	−	−	−	−	+	+	+
17	−	++	−	+	+	+	(++)	−	(++)	−	−	+	−	−	++	+	−	−	−	−	−	+	+	+
18	+	++	−	+	+	+	++	−	−	−	−	+	(++)	−	+	+	−	−	−	−	−	+	+	+
19	+	++	−	+	+	+	+	−	−	−	−	+	(++)	−	−	+	−	−	−	−	−	+	+	+
20	−	+	+	+	−	−	−	−	−	−	−	+	−	−	−	−	−	−	−	−	−	−	+	−
21	−	+	+	+	−	−	+	(+)	(+)	(+)	−	+	−	−	−	−	−	−	−	−	−	−	+	−

Table 4. The standard

PS	61C	62A	62C	62E	62F	63E	63F	64A	66B	66E	67B	67F	68C
1	—	—	—	—	—	—	—	—	—	—	—	—	++
2	—	—	—	—	—	—	+	—	—	—	—	—	+
3	—	—	—	—	—	—	+	—	—	—	—	—	—
4	—	—	—	+	—	—	+	—	—	—	—	—	—
5	—	—	—	+	—	—	++	—	—	—	+	—	—
6	—	—	—	++	—	—	++	—	—	—	+	(+)	—
7	—	—	—	++	—	—	+	—	—	—	+	(+)	—
8	—	—	—	++	—	+	+	—	—	—	+	(+)	—
9	—	—	—	++	—	+	—	—	+	(+)	+	(+)	—
10	—	(+)	—	++	+	++	—	—	+	(+)	+	(+)	—
11	+	(+)	+	+	+	++	—	(+)	++	(+)	—	(+)	—
12	—	—	—	—	—	+	—	—	++	—	—	(+)	—
13	—	—	—	—	—	+	—	—	(+)	—	—	—	—
14	—	—	—	—	—	—	—	—	(+)	—	—	—	—
15	—	—	—	—	—	—	(+)	—	(+)	—	—	—	—
16	—	—	—	—	—	+	—	—	(+)	—	—	—	—
17	—	—	—	—	—	+	—	(+)	(+)	—	—	—	—
18	—	—	—	—	—	++	—	(+)	+	(+)	—	—	—
19	—	—	—	—	—	+	—	—	++	—	—	—	—
20	—	—	—	++	—	—	—	—	++	—	—	—	—
21	—	—	—	++	—	—	—	—	++	—	—	—	—

Table 5. The standard

PS	82F	83E	84F	85D	85F1-6	86E	87A	87C	87F	88D	88EF	89B	90BC	91D
1	—	—	—	—	+	(+)	—	—	—	—	—	—	++	—
2	—	+	—	—	+	—	—	—	—	+	+	+	+	—
3	—	+	—	—	+	—	—	—	—	+	+	+	+	—
4	—	+	—	—	+	—	—	—	—	+	+	+	+	—
5	—	+	—	—	+	—	—	—	—	+	+	+	+	—
6	—	+	—	—	+	—	—	—	—	+	+	+	+	—
7	—	+	—	—	+	—	—	—	—	+	—	+	+	—
8	—	+	—	—	++	—	—	—	—	++	—	+	—	—
9	++	+	—	—	++	—	—	—	—	++	—	+	—	—
10	++	+	(+)	—	++	—	—	—	—	++	—	+	—	+
11	++	+	(+)	+	++	—	—	—	—	++	—	+	—	++
12	+	+	(+)	++	++	—	—	—	—	+	—	+	+	++
13	—	+	—	++	++	—	—	—	—	+	—	+	++	+
14	—	+	—	++	++	—	—	—	—	+	—	(+)	++	—
15	—	—	—	++	++	—	—	—	—	+	—	(+)	++	—
16	—	—	—	++	++	—	—	—	—	+	—	—	+	—
17	—	—	—	++	++	—	—	—	+	+	—	—	+	—
18	++	—	—	++	++	—	—	—	+	+	—	+	—	—
19	—	—	—	—	++	—	+	+	+	+	—	+	—	—
20	—	—	—	—	++	+	+	+	—	+	—	+	—	—
21	—	+	—	—	++	++	+	—	—	++	—	+	—	—

puff stages for 3L

69A	70C	70E	71B	71DE	72D	73B	74C	74EF	75B	75CD	76A	76D	77E	78D
(+)	—	—	—	—	—	—	—	—	—	—	—	—	—	—
—	—	—	—	+	—	—	(+)	+	+	—	—	—	—	—
—	—	—	—	+	+	—	(+)	++	++	—	—	—	—	—
—	—	—	—	+	+	—	—	++	++	—	—	—	—	—
—	+	—	—	+	+	—	—	++	++	—	—	—	—	—
—	+	—	—	+	+	—	—	++	++	—	—	—	—	+
—	+	—	—	+	+	—	—	++	++	—	—	—	(+)	++
—	+	—	—	+	+	—	—	+	+	—	—	—	(+)	++
—	+	(+)	—	++	+	—	—	+	+	—	—	—	(+)	++
—	+	+│	—	++	+	—	—	—	—	—	—	—	—	++
—	(+)	(+)	(+)	++	+	+	—	—	—	—	(+)	—	—	+
—	—	—	(+)	++	+	+	—	—	+	—	(+)	+	—	—
—	—	—	(+)	++	+	+	—	—	+	—	—	+	—	—
—	—	—	—	++	+	—	—	—	+	—	—	+	—	(+)?
—	—	—	—	++	+	—	—	—	+	—	—	—	—	(+)?
—	—	—	—	++	+	—	—	—	+	—	—	—	—	—
—	—	—	—	+	+	—	—	—	+	—	—	—	—	—
++	—	—	—	+	+	—	—	—	+	++	—	—	—	—
—	—	—	—	(+)	—	—	—	++	++	+	—	—	—	—
—	—	—	—	(+)	—	—	—	—	+	—	—	—	—	—
—	—	—	—	(+)	—	—	—	—	+	—	—	—	—	—

puff stages for 3R

92A	93B	93D	93F	95B	95D	95F	96A	96E	97C	98B	98F	99B	99D	99E	100DE
—	—	—	+	—	—	—	—	—	—	—	—	(+)	—	—	—
—	—	+	—	—	—	—	—	—	—	—	—	(+)	—	—	—
—	—	+	—	—	—	—	—	+	—	—	—	(+)	—	—	—
—	—	+	—	—	—	—	—	+	—	—	—	(+)	—	—	—
—	—	+	—	—	—	—	—	+	—	—	—	(+)	—	—	—
—	(+)	+	—	—	—	+	—	+	—	—	—	(+)	—	—	—
—	(+)	+	—	—	—	+	—	—	—	—	÷	(+)	—	—	—
—	—	+	—	(+)	(+)	+	—	—	—	—	+	(+)	—	—	—
—	⊤	+	—	(+)	(+)	+	—	—	—	—	++	(+)	—	—	—
—	⊤	+	—	(+)	(+)	+	—	—	+	(+)	++	(+)	—	—	—
—	⊤	+	—	—	—	—	—	—	+	(+)	++	(+)	—	(+)	—
—	⊤	++	—	—	—	—	—	+	—	—	—	(+)	—	(+)	—
—	⊤	++	—	—	—	—	—	+	—	—	—	(+)	—	(+)	—
(+)	—	(+)	—	—	—	—	(+)	+	—	—	—	(+)	—	+	—
(+)	—	(+)	—	—	—	—	(+)	+	—	—	—	(+)	—	+	+
—	—	(+)	—	—	—	—	(+)	—	—	—	—	(+)	—	+	++
—	—	(+)	—	—	—	(+)	(+)	—	—	—	—	(+)	(+)	+	++
—	—	(+)	—	—	—	(+)	—	—	—	—	—	+	(+)	+	++
—	—	(+)	++	—	—	—	—	++	—	—	—	++	(+)	++	—
—	—	+	++	—	—	—	—	+	+	+	++	++	—	++	—
—	—	++	++	—	—	—	—	—	+	+	++	++	—	++	—

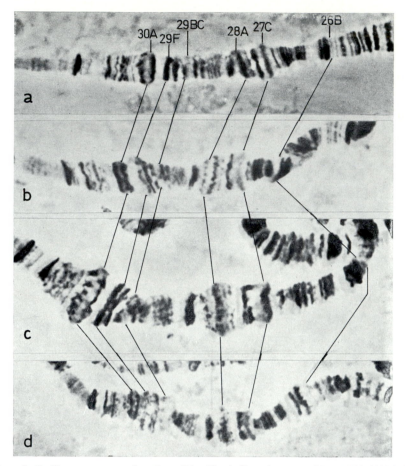

Fig. 5a—d. Puffing sequence of regions 26—30 of *D. melanogaster* 2L. a PS1 (110 h larva, Oregon), b PS14 (4 h prepupa, Oregon), c PS16 (6 h prepupa, Oregon), d PS21 (12 h prepupa *vg*6)

other puffs are uniquely active at only one time has already been mentioned. Fig. 18 emphasises the fact (Becker, 1959; Ashburner, 1967a) that the relative order of appearance and regression of groups of puffs common to both periods may differ according to the developmental stage analysed. This fact has certain implications for proposals for a cascade type of control mechanism operating on puffs.

It is to be emphasised that as a general rule all nuclei of the salivary gland of *D. melanogaster* react identically, showing qualitatively the same puffing pattern. Becker (1959) described a puff, at 15BC on the X chromosome, that was larger in proximal nuclei than in the more distal nuclei. However this puff is active in the distal nuclei and this does not reflect a qualitative difference in puffing within a salivary gland. Lytchev and Medvedev (1967) have analysed intragland variation in puff size for chromosome arm 2R of *D. melanogaster* (see also Burnett and Hartmann-Goldstein, 1971).

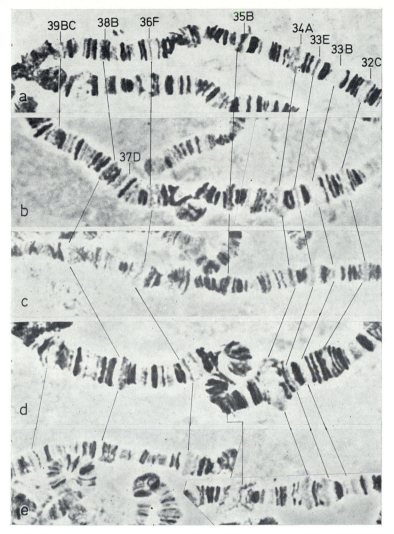

Fig. 6a—e. Puffing sequence of proximal region of *D. melanogaster* 2L. a PS1 (110 h larva, Oregon), b PS6 (115 h larva, Oregon), c PS9 (120 h larva, Oregon), d PS12 (2 h prepupa, Oregon), e PS17 (8 h prepupa, Oregon). Fig. 6e has been dissected

Although the salivary gland chromosomes of L2 and very young L3 are clearly polytene their small size precludes analysis of their puffing patterns. BECKER (1959) did note, however, puffing activity at the end of L2. Similar remarks apply to the polytene chromosomes of other larval tissues of *D. melanogaster*. For example the larval fat body (Fig. 17) malpighian tubules, gut, humeral disc associated tissue (LAMPRECHT and REMENSBERGER, 1966), epidermal cells etc. all possess thin polytene chromosomes with evident puffs. The polytene chromosomes of the ovarian nurse cells in certain female sterile mutants of *D. melanogaster* have been described by KING (KING, SANG and LETH, 1961) and may be amenable to analysis. Indeed

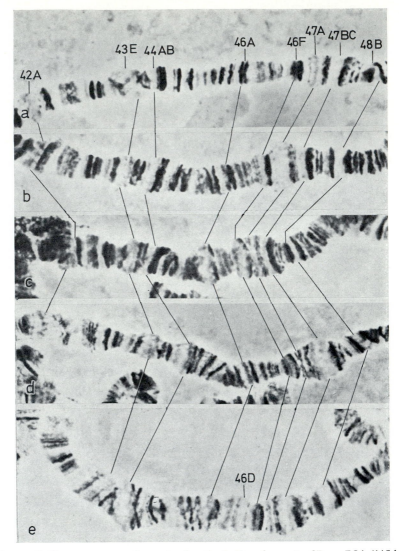

Fig. 7a—e. Puffing sequence of proximal end of *D. melanogaster* 2R. a PS4 (115 h larva, Oregon), b PS8 (120 h larva, Oregon), c PS14 (4 h prepupa, Oregon), d PS18 (10 h prepupa, *ast ho ed dp cl* stock), e PS21 (12 h prepupa, *ast ho ed dp cl* stock)

Fig. 8a—e. Puffing sequence of middle section of *D. melanogaster* 2R. a PS1 (110 h larva, *vg6*), b PS6 (115 h larva, Oregon), c PS10 (0 h prepupa, Oregon), d PS15—16 (6 h prepupa, *vg6*), e PS18 (8 h prepupa, *vg6*)

Fig. 9a—d. Puffing sequence of distal end of *D. melanogaster* 2R. a PS1 (110 h larva), b PS6 (115 h larva), c PS11 (2 h prepupa), d PS14 (4 h prepupa). All Oregon. (From Ashburner, 1967a)

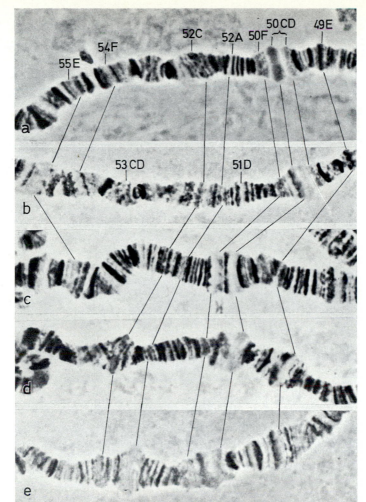

Fig. 8 a—e

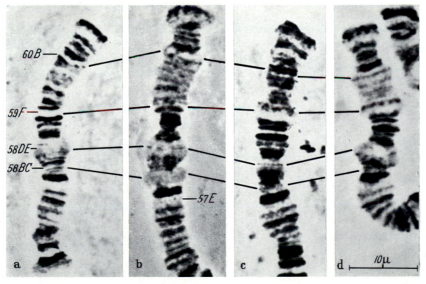

Fig. 9
a—d

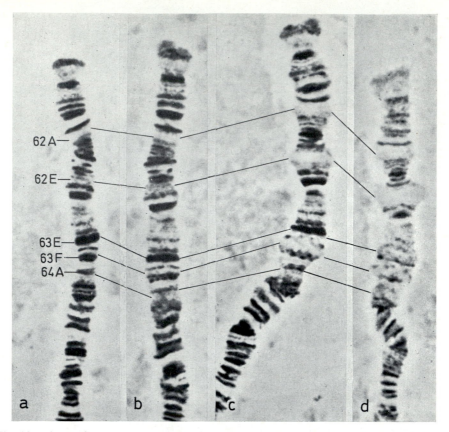

Fig. 10a—h. Puffing sequence of distal end of *D. melanogaster* 3L. a PS1 (110 h larva), *h eyg cp/mot³⁶ᵉ* F1) b PS2 (110 h larva, Oregon), c PS6 (115 h larva, *vg6/ast ho ed dp cl* F1), d PS8 (120 h larva, *ft*)

SCHULTZ (1965) has already reported the results of preliminary work on these chromosomes. Quite remarkable is the condition of the X chromosome which, to quote SCHULTZ is "completely diffuse, a sort of giant puff" in some nurse cell nuclei of the mutant *fs(2)B*.

III. Genetic Variation in Puffing

The great attraction of *Drosophila melanogaster* for studies on puffing is the existence of a very large number of well-defined mutants and of the genetical techniques for their manipulation. Hopes that these would enable correlations between puffs and phenotypes to be made have not yet been fulfilled. However the detailed descriptions of standard puffing patterns are absolutely essential for any study of variation, be it genetic or environmentally caused, in puffing. The standardisation of a "wild type" puffing pattern has led to the recognition of various abnormal puffing patterns in particular stocks of *D. melanogaster*.

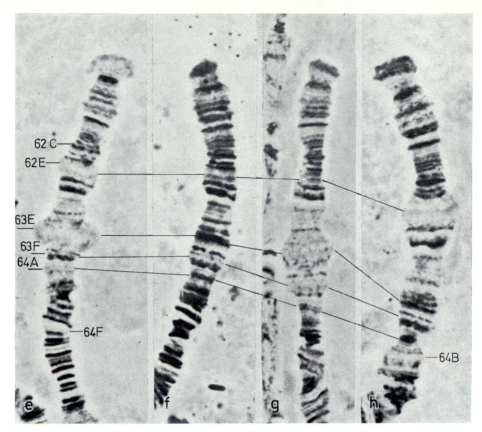

Fig. 10.   e PS11 (0 h prepupa, Barton), f PS14 (4 h prepupa, Oregon), g PS18 (8 h prepupa, *ast ho ed dp cl*). h PS21 (12 h prepupa, *ast ho ed dp cl*)

Broadly speaking variation in puffing patterns is of two types, previously described as quantitative variation and qualitative variation (ASHBURNER, 1969c, d). It must be emphasised that the "standard puffing pattern" is an arbitrarily defined phenotype and serves only for the convenience of the experimenter. Departures from this pattern will, where there is at least reasonable evidence for supposing that these are due to particular genetical constitutions, be referred to as mutant patterns.

A. Quantitative Variation

This term includes examples of particular puffs differing in either their size and/or the time of their activity from the standard pattern. Several examples of variation in both puff size and puff timing were discovered during detailed comparisons of the puffing patterns of two stocks of *D. melanogaster* (Oregon-R and vg6) and of *D. melanogaster* and its sibling species *D. simulans* (see section IV).

Some examples of each type of variation are given in Figs. 19 and 20. The analysis of the puffing patterns of the vg6 stock was extended to the puffing patterns

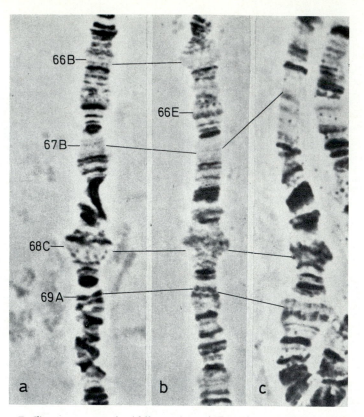

Fig. 11a—c. Puffing sequence of middle section of *D. melanogaster* 3L. a PS1 (110 h larva, *h eyg cp/mot^{36e}* F1), b PS11 (0 h prepupa, Oregon), c PS18 (8 h prepupa, Oregon)

of the vg6/Oregon-R F1 hybrids. Puffs, differing in their behaviour in the parental strains showed one of four types of behaviour in the F1: (1) similar to maternal stock (vg6), (2) similar to paternal stock (Oregon-R), (3) intermediate between parental stocks or (4) a novel pattern. Such variation in the activity of particular puffs will doubtless be found to be ubiquitous, although information as to its extent is meagre. Apart from the studies of Russian authors, to be mentioned below, little attempt has been made to assess the extent of variation in puffing in natural populations of *Drosophila*.

Kiknadze (1966) has studied the puffs of various stocks derived from wild caught *D. melanogaster* and their yellow mutant sublines. She found differences in activity at several loci, some (e.g. 4AB and 17E on the X chromosome) apparently correlated with the yellow mutation. Lytchev (1965) analysed the puffs of two wild type stocks of *D. melanogaster*, Novosibirsk and Batumi, and of an inbred line derived from the Batumi population. Many differences in puffing were observed in this study. Of particular interest was the conclusion that inbreeding led to a general depression of activity, especially on the X chromosome. Later Lytchev and Medvedev (1967) studied variation in puffing on 2R within salivary glands and

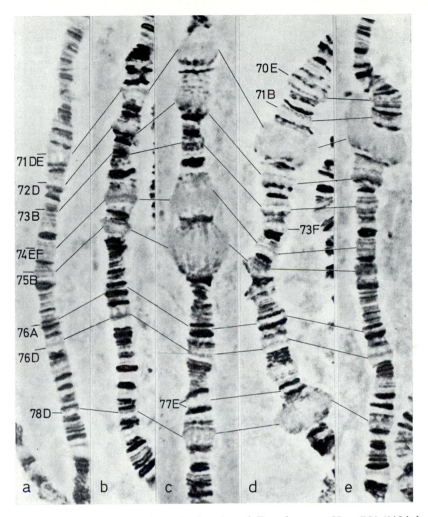

Fig. 12a—g. Puffing sequence of proximal region of *D. melanogaster* 3L. a PS1 (110 h larva, Oregon/Pacific F2), b PS2—3 (110 h larva, Oregon), c PS6 (115 h larva, Oregon), d PS10 (0 h prepupa, *h eyg cp/mot³⁶ᵉ* F1), e PS14 (4 h prepupa, Barton)

between individuals of the inbred Batumi stock. However all three of these studies have apparently failed to appreciate the rapid changes occurring in puffing activity towards the end of L3 and, although interesting, many of their conclusions may result from the developmental heterogeneity, rather than the genetic heterogeneity, of their material.

Understanding of the significance of quantitative variation in puffing activity is hindered since (a) we lack knowledge of the basic molecular factors involved in size or timing variation, and (b) the difficulty of further genetic analysis of these types of variation.

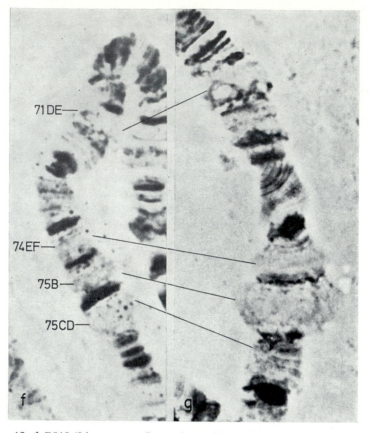

Fig. 12. f PS18 (8 h prepupa, Oregon), g PS19 (12 h prepupa, *ast ho ed dp cl*)

B. Qualitative Variation

In this subsection two classes of variation in puffing observed in *D. melanogaster* will be considered. Although they differ greatly from each other they will be treated together since both result from single allelic substitution.

The first class is exemplified by the mutant lethal-giant-larva (*l(2)gl*) of BRIDGES. Homozygous lethal-giant-larvae fail to form normal puparia although they do develop to the third instar. Often these larvae continue to live for many days beyond the time that their heterozygous sibs have formed puparia; they become rather bloated and transparent in appearance and die either as over aged larvae or following an abnormal tanning of their cuticle (the "pseudopupae" of HADORN, 1937). The *l (2)gl* phenotype is thought to result from a lack of ecdysone during the L3 and some experiments (e.g. SCHARRER and HADORN, 1938; KARLSON and HANSER, 1952) tend to confirm this view. However alternative explanations, for example a failure of the neurosecretory cells of the brain to activate the prothoracic gland, have not been investigated and cannot be ruled out. All larval tissues, including the salivary gland (see GROB, 1952) are considerably hypotrophied in *l (2)gl* homozygous L3 and

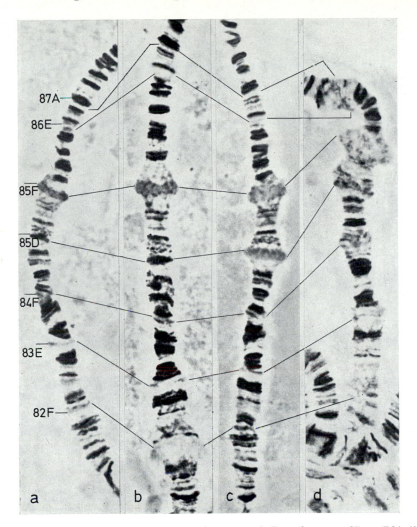

Fig. 13a—d. Puffing sequence of proximal region of *D. melanogaster* 3R. a PS1 (110 h larva, Oregon), b PS10 (0 h prepupa, Oregon), c PS13 (4 h prepupa, Oregon), d PS21 (12 h prepupa, *ast ho ed dp cl*)

the nuclear DNA content of salivary gland nuclei is lower than normal, (WELCH, 1957). This last point is reflected in the very low polyteny of their salivary gland chromosomes, (Fig. 21), a fact making their study difficult. BECKER (1959) discovered, and ASHBURNER later confirmed, that the salivary gland chromosomes of *l (2)gl* L3 or "pseudopupae" completely fail to initiate the normal puffing sequence characteristic of larval/prepupal development. Very few puffs are seen in these animals, and at least one of those that is active is not normal for development, (3L : 63BC). Puffing in the *l (2)gl/+* heterozygotes is normal and both heterozygous and homozygous larvae react to a temperature shock in a manner very similar to that of wild

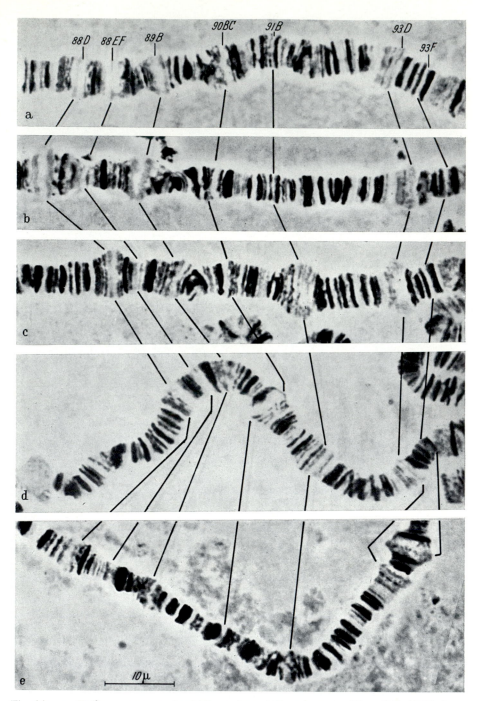

Fig. 14a—e. Puffing sequence of middle region of *D. melanogaster* 3R. a PS1 (110 h larva, b PS9 (0 h prepupa), c PS11 (1/2 h prepupa), d PS14 (4 h prepupa), e PS20 (12 h prepupa), All Oregon. (From Ashburner, 1967a)

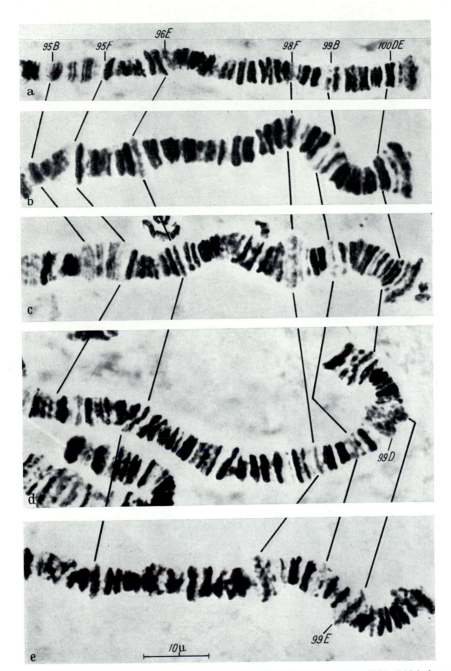

Fig. 15a—e. Puffing sequence of distal region of *D. melanogaster* 3R. a PS1 (110 h larva), b PS6 (115 h larva), c PS9 (120 h larva), d PS17 (8 h prepupa), e PS20 (12 h prepupa). All Oregon. (From Ashburner, 1967a)

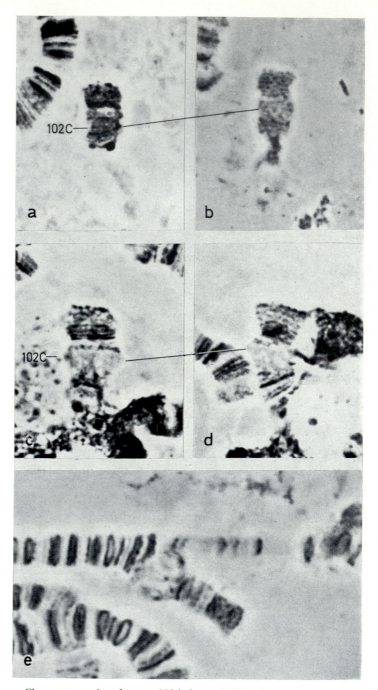

Fig. 16a—e. Chromosome 4. a from a 110 h larva (PS2/3, Oregon), b from a 120 h larva (PS8/9, $Dp(1:f)R)/y\ dor^1/y/dor^1$), c and d from 12 h prepupae (PS21, *ast ho ed dp cl*), with puff at 102C, e Chromosome 4 in T(3 : 4)83 with puff at 4 : 101—102A. Photograph supplied by Roberts

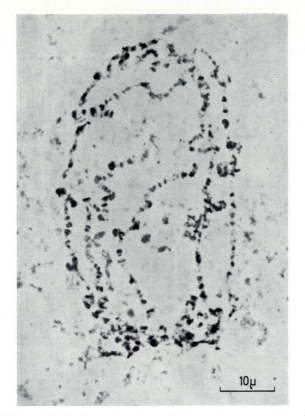

Fig. 17. Polytene fat body nucleus from L3 just prior to puparium formation

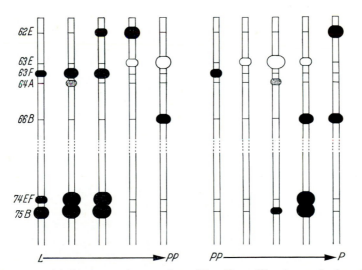

Fig. 18. Sequence of induction and regression of 7 puffs on 3L prior to the larval/prepupal (L/PP) and prepupal/pupal (PP/P) moults (From Ashburner, 1967a)

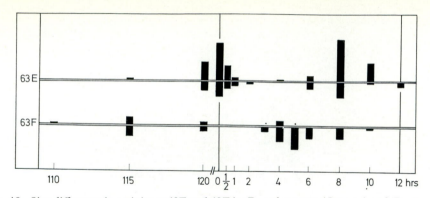

Fig. 19. Size difference in activity at 63E and 63F in *D. melanogaster* (Oregon) and *D. simulans* (Berkeley). The *D. simulans* histograms are inverted below the relevant *D. melanogaster* histograms. (From ASHBURNER, 1969b)

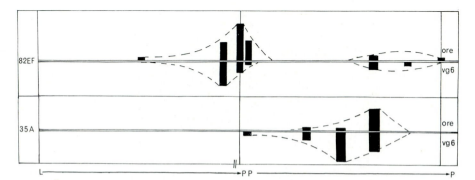

Fig. 20. Timing difference in activity at 82F (=82EF) and 35B (=35A) in the Oregon and *vg6* stocks of *D. melanogaster* (From ASHBURNER, 1969e)

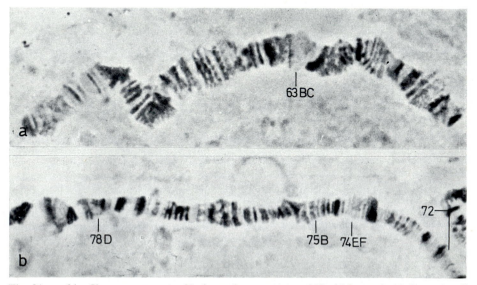

Fig. 21a and b. Chromosome arm 3L from homozygous *l*(2)*gl* L3 aged 11 days. Small puff at 63BC in a and absence of puffing at 72, 74, 75 or 78 in b

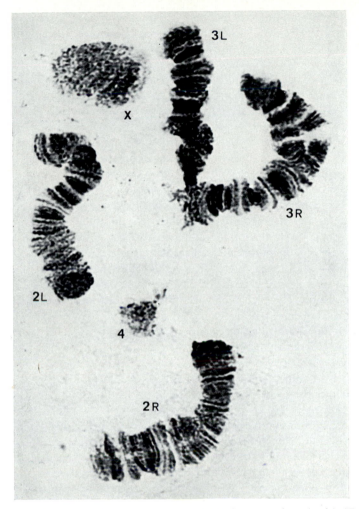

Fig. 22. Abnormal salivary gland chromosomes of *lethal tumorous larva* (male). (From Kobel and van Breugel, 1967)

type animals with characteristic puff induction (see section VI.B and Ashburner, 1970c).

Another larval mutant analysed cytologically is *l*(2) tumorous larvae (*l* (2)*tl*) of Kobel and van Breugel (1967). This lethal, isolated from Gardener's tumorous head (*tu-h*) stock studied by Rodman and Kopac (1964) and by Rodman (1964) is, like *l* (2)*gl*, characterised by a failure of puparium formation and by an extended larval survival. Unlike *l* (2)*gl*, however, the larval tissues and imaginal discs of *l* (2)*tl* are not grossly hypotrophied although they are clearly abnormal (Kobel and van Breugel, 1968). Homozygous *l* (2)*tl* L3 develop melanotic "pseudotumours". The salivary gland appears to break down prematurely and there is an encapsulation by haemocytes (with their subsequent melanization) of salivary gland cells. The salivary gland chromosomes of old *tu-h* L3 are abnormal (Rodman and Kopac, 1964; Rodman, 1964); they become much wider and shorter than normal, a transformation seen in its extreme form in the hemizygous X chromosomes of male larvae (Fig. 22). Rodman (1964) initially reported the failure of puff formation on 3L in certain *tu-h* sublines. Subsequently in a study of *l (2)tl*, isolated from *tu-h* and showing the same cytological abnormalities as *tu-h*, van Breugel and Bos (1969) found that very old (13—30 day) *l (2)tl* larvae developed puffs on 3R normally associated with temperature shock (i.e. 87A, 87B, 93D and 95D). Other puffs, normal to development (e.g. 88D, 88EF, 85F) were found to be active in these larvae but further information is needed to justify the authors' conclusion that the "puffing pattern of *ltl* may be more accurately related to the situation in the very late third instar larvae in normal development".

A third larval lethal studied is *dor^l* (previously known as 1(1)7). Hemizygous *dor^l* males die as L3 with extensive melanotic pseudotumour formation and other abnormalities (e.g. occlusion of the mid-gut (Russell 1940)). Cytogenetically the *dor* locus lies within, or immediately adjacent to, the region of the X chromosome puff 2B5.6 (Rayle and Hoar, 1969). This puff is active in *dor^l* male L3 but following the regression of puffs characteristic of the early puff stages (e.g. 3C11.12) no further changes in puffing occur in this mutant until just prior to death (several days later) when there is abnormal puffing at 50CD, 50F and at a few other sites (Rayle, personal communication). Puffing in another mutant mapping in or very close to the 2B5.6 region, the mutant *halfway* of Rayle (1967), is rather different from that described for *dor^l*. Male larvae hemizygous for *hfw* typically form abnormal puparia characterised by a tanned anterior region and an untanned, larval, posterior region. The changes in puffing pattern characteristic of the late L3 of wild type occur almost normally in *hfw* males. In fact puffing is normal up to PS14 and then arrest occurs. No further changes in puffing pattern after PS14 were observed by Rayle although the animals live and the chromosomes remain analysable for almost a day further (Rayle, personal communication).

Not all larval recessive lethals with late L3/prepupal critical periods are abnormal in their puffing phenotypes. Two examined by Ashburner (unpublished) appear to have essentially normal puffing patterns. These mutants are *lethal(2)histolytic* of Thompson and *lethal(2) giant discs* of Bryant. The *l (2)hist* mutant when homozygous results in death after puparium formation apparently accompanied by a sudden and almost complete histolysis of all tissues. The *l (2)gd* homozygous larvae form normal puparia and die following the L4/pupal ecdysis. This mutant has hypertrophied imaginal discs.

In contrast to these lethal mutants there are also a number of genotypes which result in the loss, or addition, of a specific puff relative to the standard pattern, while the activity of all other puffs of the pattern remains unchanged. In a few cases analysis has demonstrated that this 'mutant' phenotype results from a single allelic substitution. Several 'mutant' puff phenotypes have been described although in no case has it yet been possible to correlate, without ambiguity, a mutant puff condition with an abnormal phenotype at some other level of observation or analysis.

A list of the single puff mutants known from *D. melanogaster* and *D. simulans* is given in the accompanying table (Table 6). The genetic and cytogenetic behaviour of many of these puffs has been discussed in other publications (Ashburner, 1969 c, d;

Table 6. Putative "puff mutants" in *Drosophila melanogaster*

Locus	Mutant stock	Mendelian segregation demonstrated?	Cytological behaviour in hybrids	Reference
X:2B	*halfway* males lacks 2B 13—17	—	Puffing at 2B 5—6 and 2B 13—17 of *hfw* homologue lags behind + homologue in heterozygous females, when asynapsed.	RAYLE (1967)
4AB, 17E	Additional puffs in yellow mutant lines	—	—	KIKNADZE (1966)
4F1—4 7B1—3	Additional puffs in *D. simulans* Berkeley	F1 sterile	—	ASHBURNER (1969a)
2:22B4—5	Oregon and vg6 lack	yes	Homozygous synapsed and asynapsed	ASHBURNER (1970d)
22B8—9	Additional puff in *ast ho ed dp cl*	yes	Homozygous synapsed/heterozygous asynapsed	ASHBURNER (1970d)
24E	Additional puff in *fat*	—	—	SLIZINSKI (1964)
46A	*D. simulans* Berkeley lacks	F1 sterile	Heterozygous	ASHBURNER (1969b)
3:64C	Additional puff in vg6 and related lines	yes	Homozygous synapsed/heterozygous asynapsed	ASHBURNER (1967a, 1967b, 1969c)
4:101—102A	Additional puff in *T(3:4)83*	—	—	ROBERTS (1969, 1970)

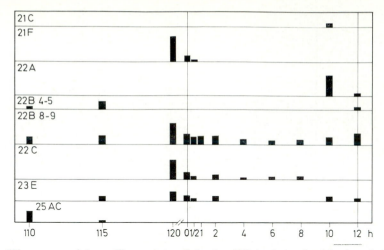

Fig. 23. Histograms of the puffing activity of the tip of 2L in the *ast ho ed dp cl* stock.Compare with Fig. 1 of Ashburner 1967a. Note the activity of the two puffs in section 22B (22B4—5 and 22B8—9) not active in Oregon

1970d), and this information is only summarised here. Six of these puff mutants, are puffs supplementary to the standard pattern (X : 4AB, 17E, 2L : 22B8—9, (Figs. 23 and 24) 3L : 64C (Fig. 25) and 4 : 101—102 (Fig. 16e)). Some of these puffs at least, are developmentally specific.

The puff 24E was described from the mutant *fat* of *D. melanogaster* and is close to the cytogenetic position of this mutant on the salivary gland chromosome map (Slizynski, 1964). Unfortunately this puff has not been subjected to a genetic analysis and its low and variable expressivity (very dependent upon larval culture conditions) makes its study rather difficult. Nevertheless the reported correlation between the presence of this puff and cytoplasmic vacuoles in the salivary gland cells (Slizynski, 1964; see Chaudhuin, 1969) should be followed up.

C. Variation Resulting from Chromosome Aberrations

A very large number of chromosome aberrations, inversions, translocations, deletions etc., are known in *D. melanogaster*. Even some very complex chromosomes, such as, multiple, balancer chromosomes, have been found to have patterns of puffing activity essentially similar to the standard pattern.

Nevertheless the phenomenon of position effect variation in *Drosophila* (Lewis, 1950; Baker, 1968) has led to a search for comparable phenomena at the puff level, in particular for the suppression of a puff as the result of its translocation close to a novel euchromatic: heterochromatic break. There are a few instances of such an effect. The X : 4 translocation $T(1 : 4) w^{258-21}$ transfers the tip of the X to region 3E to chromosome four and, when heterozygous with a recessive white mutant, produces a position effect at the white and Notch loci. The break in the X chromosome is just proximal to the puffs 3C and 3E and in translocation heterozygotes there is suppression of puffing at these loci on the 4 : X chromosomes as the result

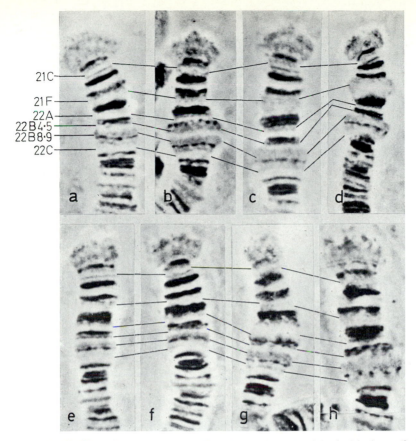

21C—
21F—
22A—
22B4·5—
22B8·9—
22C—

a b c d

e f g h

Fig. 24a—h. Puffing of the region 21—22 in the stock *ast ho ed dp cl* with the puffs 22B4—5 and 22B8—9 (compare with Fig. 3). a PS1 (110h larva), b PS3 (110h larva), c PS7 (115h larva), d PS9 (120 h larva), e PS14 (4 h prepupa), f PS16 (6 h prepupa), g PS18 (10 h prepupa) h PS21 (12 h prepupa)

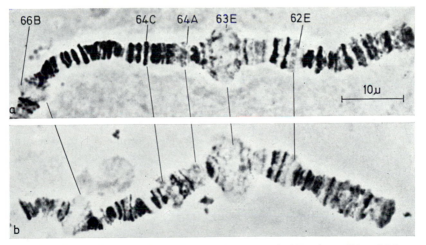

66B 64C 64A 63E 62E

10μ

a

b

Fig. 25a and b. The puff mutant 64C. a Oregon 0 h prepupa lacking the puff, b *vg6* 0 h prepupa with 64C puff active

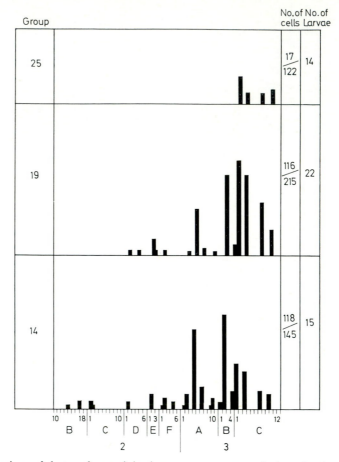

Fig. 26. Relation of heterochromatinization to temperature during development. Each block of the histogram represents the number of nuclei exhibiting heterochromatinization up to and including the chromosomal position indicated on the abscissa. The nuclei are divided into groups representing the temperature during development (25°, 19° or 14° C). The second-last column gives the number of nuclei showing heterochromatinization, over the total number examined in each group. The last column gives the number of individuals from which the nuclei were taken. A clear-cut correlation is seen between temperature and heterochromatinization. (From Hartmann-Goldstein, 1967)

of compaction (heterochromatinization) of this chromosome region. The degree of puff suppression parallels mutant expression in flies exhibiting position effect variegation. That is to say, there is less puffing at 3CD at 18° C than at 25° C, while at both temperatures addition of an extra Y chromosome to the genome increases the degree of puffing (Rudkin, 1965; Schultz, 1965).

The cytological behaviour of this translocation was further studied by Hartmann-Goldstein (1967). She grew $T(1:4)w^{258-21}/In(1)\ dl^{49}w\ lz$ heterozygous larvae at three different temperatures (14°, 19° and 25° C). Not only did the number of nuclei showing compaction of the translocated X chromosome increase with a decrease

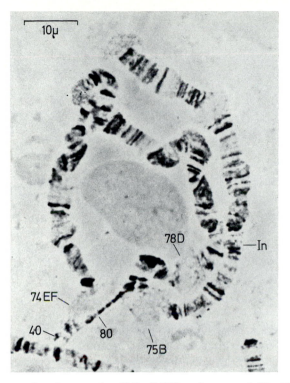

Fig. 27. Heterozygote for the complex T(2:3)C65 and a normal 3L. This chromosome is a translocation between 2 and 3, each broken in centric heterochromatin, with a second break on 3L at 75A between the puff pair 74EF and 75B. In addition there is an inversion on 3L with break points 77A and 64C. The complex thus has the new order: 2R tip to 40/75A — 74 EF — 64C/77A — 80/75A — 75B — 77A/64C — tip 3L. In this preparation this complex is synapsed with a normal homologue, the inversion cross is marked In, the centric heterochromatin of 3 lying distal to 75B is marked 80, and the centric heterochromatin of 2 proximal to 74EF is marked 40. Chromosome arm 2R is off the photograph

in culture temperature but so did the actual extent of compaction. At 25° C the most distal limit of compaction was within region 3B, at 19° C within region 2D while at 14° C it was the band 2B13, a distance of 65 bands from the break point (Fig. 26). Temperature shift experiments established that the major critical period for temperature to affect the extent and frequency of compaction was during the early embryonic period. Under different temperature regimes the extent of variegation for both white (malpighian tubules) and Notch (wings) paralleled the extent of compaction of the translocated X in salivary gland chromosomes.

The coordinate behaviour of adjacent puffs during development has often been noted (BECKER, 1959; ASHBURNER, 1967a). A very clear example is the pair of puffs on 3L at 74EF and 75B. These are among the first puffs to respond to an increase in ecdysone titre either normally during development, or as the result of ecdysone treatment (see section V). The physical proximity of these two puffs is not necessary for their coordinate behaviour. This is illustrated by their behaviour in heterozygotes

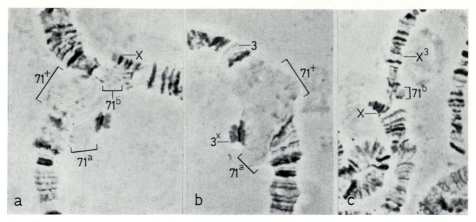

Fig. 28a—c. Heterozygotes for the reciprocal translocation $T(1:3)\mathrm{sc}^{260-15}$. The break in 3L splits the puff 71DE into two puffs on the $3L^{x}(71^{a})$ and $X^{3L}(71^{b})$ chromosomes. In a the chromosomes are synapsed (71^{+} is the intact 71DE puff of the 3L normal homologue). b and c show, asynapsed, the $3L^{x}/3L$ and X^{3L}/X chromosomes respectively with their 71 "subpuffs"

for the complex $T(2:3)C65$ of Roberts. This translocation separates 74EF and 75B and places both puffs close to centric heterochromatin (Fig. 27). Yet the developmental behaviour of both puffs is unaltered as the result of their novel positions. Ellgaard and Brosseau (1969) have studied, in a similar manner, effects of translocations on the activity of the adjacent puff pair 87A and 87B. During normal development the activity of these puffs is not coordinate; in fact only 87A shows significant puffing at any period studied. Yet after a temperature shock (see section VI.B) large puffs are regularly and apparently coordinately induced at these two sites. A number of translocations were studied by these authors for their effects on the temperature shock induced puffs at 87A and 87B. Chromosome breakage distal, proximal or between the puff pair in no case altered the behaviour of these puffs in either the translocated or wild type chromosomes of heterozygous larvae.

Breakage of a chromosome within a puff region, rather than adjacent to it, has been observed in one or two instances in *D. melanogaster*. For example the reciprocal $T(1:3)sc^{260-15}$ has 3L broken within the very large puff 71DE. As shown in figure 28 each region now organises a separate puff; during development the behaviour of these two "new" puffs and of the intact 71DE puff in heterozygotes is coordinate. Until many more breaks are studied in this puff region it is difficult to know whether this result is due to redundancy within the puff region or to a functional complexity of the puff itself.

Duplication of chromosome regions allows the study of puffing in regions of abnormal dosage. Only one example has been studied in detail and this is the free X chromosome fragment $Dp(1:f)R$ derived from the ring chromosome $R(1)2$ and showing position effect variegation at several loci, notably at *dor*. This duplication consists of the tip of the X chromosome (in fact the terminal bands are lacking) to region 3A attached to an X centromere. It thus includes several puffs, e. g. 2B5—6, 2B13—17, and 2EF. In Dp carrying males the duplication often synapses,

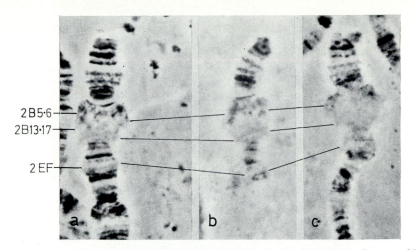

2B5·6

2B13·17

2 EF

a   b   c

Fig. 29a—c. The duplication Dp(1 : f)R in the stock $Dp(1 : f)R/y\,dor^e/y\,dor^e$. a and b from a single female nucleus (PS8) with the synapsed X chromosomes a and the free duplication b, c from a male nucleus (PS8/9) with the duplication synapsed to the hemizygous X

more or less completely, with the normal X while in Dp females the duplication always lies away from the normal X chromosomes and is normally to be found attached to the chromocentre. In either case the pattern of puffs on the duplicated fragment has been found to be identical to that on the normal X homologues in the same nucleus (Fig. 29). The puffing patterns of the duplication and the normal X homologues are like the standard pattern (see also discussion of KORGE's experiments p. 144).

IV. *Drosophila simulans* with Notes on *D. ananassae* and *D. takahashii*

These three species are the only other members of the melanogaster species group to have been analysed with regard to their polytene chromosome puffing activity. *D. yakuba*, another member of the *melanogaster* subgroup (*melanogaster*, *simulans* and *yakuba*) awaits cytological investigation.

D. simulans is morphologically very similar to *D. melanogaster* and the two sibling species will hybridize. However all regular hybrid progeny are sterile and progenies are often of one sex only; usually that of the *melanogaster* parent of the cross. The other two species are more distinct, morphologically, from *D. melanogaster*. *D. takahashii* has been crossed with *D. melanogaster* but the hybrid larvae died within the first instar (Sturtevant, personal communication). Hybridization between *D. ananassae* and *D. melanogaster* has not been successful. The increasing reproductive isolation of *simulans*, *takahashii* and *ananassae* from *melanogaster* is reflected in their mitotic metaphase and polytene chromosome karyotypes (PATTERSON and STONE, 1952). The polytene chromosome banding patterns of *melanogaster* and *simulans* are very obviously similar (PATAU, 1935; KERKIS, 1936; 1937; HORTON, 1939; ASHBURNER, 1969b); examination of the chromosomes in the species hybrid unequivocally demonstrates their homology. The chromosomes of *simulans* differ from those of

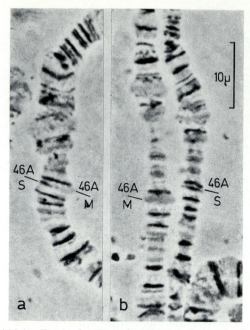

Fig. 30a and b. Puff 46A in *D. simulans*/Berkeley/*D. melanogaster* (*vg6*) hybrids. Two nuclei from the same 4 h prepupal gland with a homologues synapsed and a heterozygous puff and b homologues asynapsed with no activity on the *simulans* chromosome(s)

melanogaster by one major inversion on 3R, two small inversions near the centromeres of chromosomes 2 and 3 and other minor differences affecting few bands each.

The banding pattern of *D. takahashii* differs from that of *D. melanogaster* although there are considerable regions of great similarity. Superficially at least the polytene chromosomes of *D. melanogaster* and *D. ananassae* bear little resemblance, although there are some chromosome regions with similar banding patterns.

Puffing patterns in the Berkeley wild strain of *D. simulans* were studied in detail by ASHBURNER (1969a, b). The similarity with the standard *D. melanogaster* pattern was very obvious; indeed the extent of the differences were of a similar magnitude to those distinguishing two different stocks of *D. melanogaster* itself (ASHBURNER, 1969e). Some of the differences observed between *D. melanogaster* and *D. simulans* have already been discussed. It is difficult to know just to what extent these differences reflect species differences until further stocks of *D. simulans* have been studied. The absence of the puff 46A in *D. simulans* Berkeley (Fig. 30) is not a species specific difference as this puff is active in some, but not all, other strains of *D. simulans* examined (active in strain Guatemala-3 but not in strain South Africa).

Although a detailed study of puffing in *D. takahashii* is awaited a preliminary analysis by the author of chromosome arm 3L demonstrates that despite the structural changes undergone in this chromosome arm many of the features of the late larval puffing cycle remain clearly recognisable.

The puffing patterns of *D. ananassae* have been studied by MORIWAKI and ITO (1969). These authors analysed in detail the developmental behaviour of puffs on chromosome arms 2L and 2R (homologues of 3R and 3L respectively of *D. melano-gaster*) in two homokaryotic strains (AA and BB) and in their (inversion) hybrid. The activity of the puffs studied was independent of the chromosome polymorphism. As in other species studied there is a peak of puffing activity in *D. ananassae* at the time of puparium formation. Most puffs are active only at this time although a few (e.g. 60B) are specific for younger L3 ages. Despite the fact that the rate of develop-ment of the AA strain was slower than that of either the BB or the AB hybrid strains the activity of the puffs in each strain was very closely related to the develop-mental, rather than chronological, age of the larvae. No information on puffing in later prepupal development has yet been published for this species. The number of puffs recorded by MORIWAKI and ITO (48) is certainly an underestimate of the actual number of active sites in this species.

V. Hormonal Control

The dramatic changes in puffing activity in L3 preceding puparium formation result from an increase in the ecdysone titre of the haemolymph at this period. This was demonstrated by BECKER (1962). Ligature of L3 behind the brain-ring gland complex prior to the critical period (FRAENKEL, 1935) prevents puparium formation posterior to the ligature since the tissues there are separated from the source of the hormone ecdysone. Ligature also divides the salivary gland into two regions and BECKER analysed the puffing patterns in salivary glands from ligatured larvae. No difference in puffing was found when the ligature was applied after the release of ecdysone. In animals ligatured prior to this event only those nuclei in cells anterior to the ligature proceeded with the characteristic late L3 puffing cycle, those posterior to the ligature remaining at early PS. Transplantation experiments (see p. 144) confirmed the induction of the post PS1 puffing patterns by ecdysone.

Further investigations of the induction of puffs in *D. melanogaster* has proceeded by studies of the chromosomal response following injection of ecdysone, paralleling previous studies in *Chironomus* (CLEVER, 1961) and in *D. hydei* (BERENDES, 1968), and by studies of induction in salivary glands cultured *in vitro* (ASHBURNER, un-published).

Injection of $10^{-4}\mu g$ ecdysone into 90 h L3 (i.e. 20—30 h before the normal initiation of the pre-puparium formation changes in puffing) induces puffs at 74EF, 75B, 23E and 2B within fifteen minutes. Large puffs characteristic of PS 1 i.e. 25AC and 68C regress. Within 30 min all injected animals are at PS 2—3. Longer term experiments at a higher concentration ($10^{-3}\mu g$ injected) result in the following puff stages at various times after injection: PS 2—3 (1 h), PS 3—5 (2 h), PS 4—7 (4 h), PS 9—10 (6 h), PS 10—12 (8 h) and PS 13—14 (10 h). Very similar results were obtained with 20—OH ecdysone. The sequence of changes in puffing pattern induced by ecdysone injection into 90 hr L3 was identical to that normally seen during development. However the correlation between specific puff stages and specific morphological events was uncoupled in these experimental animals. Usually puparium formation occurs at PS 10—11, in fact in non-mutant stocks this association is invariable. In the injected animals PS 10 patterns were seen about 6 hrs after injection yet all of the animals remained active L3. No acceleration of puparium formation occurred as a result of ecdysone injection in these experiments and even the

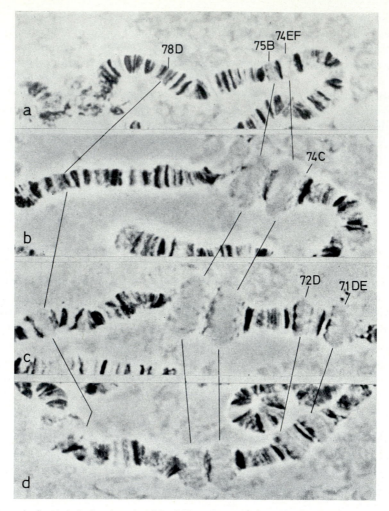

Fig. 31a—d. *In vitro* induction by 20-OH ecdysone of puffs on 3L of *D. melanogaster* (Canton-S). Glands were explanted at PS1 into Grace's insect tissue culture medium containing 3.5×10^{-4}M 20-OH ecdysone (Zoecon) and incubated in depression slides at 21° C. a after 1 h (PS1—2), b after 2 h (PS4), c after 4 h (PS6), d after 6 h (PS7—8)

PS 13—14 animals were larvae. In normal development this PS occurs in 2—4 h prepupae.

Salivary glands of L3 cultured *in vitro* in Grace's insect tissue culture medium respond to ecdysone. At relatively high concentrations (3.54×10^{-4}M) the response of PS 1 glands to ecdysone, 20—OH ecdysone, or inokosterone is very similar to that seen *in vivo*, although the reaction of a few puffs (notably 63F) may not be as large (Fig. 31). *In vitro* the puffing patterns proceed through their normal cycle up to PS 13 which is reached after 20 hr culture at 20° C. At lower concentrations of ecdysone (e.g. 3.54×10^{-5}M) the pattern proceeds to PS 2—3, and then regresses-

In spite of recent reports of a modification in puffing by juvenile hormone in *Chironomus* (LAUFER and GREENWOOD, 1969; LEZZI and GILBERT, 1969; LAUFER and HOLT, 1970) no effect of the LAW and WILLIAMS JH analogue on puffing in *Drosophila melanogaster* has so far been detected. Third instar larvae of varying age (from 10 h to 40 min before puparium formation) have been topically treated with this analogue and both the puparium formation and pupation cycles of puffing analysed. Topical treatment of L3 in this way results in a disturbance of imaginal differentation of the abdomen (ASHBURNER, 1970b).

VI. Experimental Modification of Puffing

In addition to responding to altering hormone environments puffs also respond to a variety of other environmental factors.

A. Nutritional Factors

Two studies (BURNET and HARTMANN-GOLDSTEIN, 1971; LYTCHEV, 1965) have attempted to discover whether varying the quality of the food available to growing larvae will produce changes in the puffing patterns. BURNET and HARTMANN-GOLDSTEIN cultured larvae on live yeast medium and on an RNA deficient medium. LYTCHEV grew two different strains of *D. melanogaster*, the Batumi population and an inbred line derived from it, on both high and low protein media. LYTCHEV observed very little effect of diet on puffing. In BURNET and HARTMANN-GOLDSTEIN's study some changes in the timing of puffing activity at specific loci, in particular earlier activity in larvae grown on the RNA deficient medium, were attributed to nutritional differences.

B. Temperature

Two different factors need consideration here. First, the effect on puffing of continuous culture at different temperatures, and second the effect of brief temperature shocks disturbing an otherwise constant culture temperature.

There has been little study of puffing patterns of *Drosophila* cultured continuously at different temperatures. We may infer from the figures of SCHULTZ (1965) that there will be some perturbation of the patterns as the result of different temperature regimes (see also BURNET and HARTMANN-GOLDSTEIN, 1971).

Temperature shock effects on puffing have been analysed many times following the initial observation of RITOSSA (1962, 1963) that transfer of larvae from 25° to 37° C for 40 min or so resulted in the dramatic induction of relatively few specific puffs. The reaction of *D. melanogaster* to such temperature shocks has been further analysed by ASHBURNER (1970c). Transfer of L3 or prepupae from 25° to 37° C results, within five minutes, in the induction of puffs at the following loci: 2L:33B; 3L: 63BC, 64F, 67B, 70A; 3R: 87A, 87B, 93D and 95D (Fig. 32). This reaction occurs uniformly in all stocks of *D. melanogaster* (and *D. simulans*) examined and is almost identical whether the treated animals are 110 h L3, 0 h or 4 h prepupae. In some experiments a 10th site is induced 2R: 48E after temperature shock. If the animals are kept at the high temperature for long periods the induced puffs regress. The size of the induced puffs, and the time they remain active increases with an increase in the temperature differential, at least up to 37.5° C.

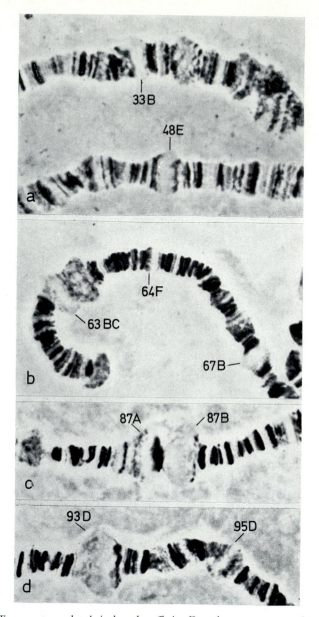

Fig. 32a—d. Temperature shock induced puffs in *D. melanogaster*, as result of 40′ at 37° C.
a 2L and 2R with 33B and 48E (*vg6* 0 h prepupa) b 3L with 63BC, 64F and 67B (Pacific,
0 h prepupa) c and d 3R with 87A, 87B, 93D and 95D (*vg6*, 0 h prepupa)

Apart from these puffs induced as the result of a temperature shock all other
puffs active at the time of commencement of the treatment, rapidly regress at high
temperature (37.5° C). On return to the normal culture temperature these developmentally normal puffs reappear although, at least after 4 h at 37.5° C, they do so

in abnormal combinations. There is also evidence (ASHBURNER, 1970c) that if at the time of temperature puffs normal to development are regressing then the rate of their regression will be slowed during the temperature shock.

C. Anaerobiosis

Deprival of *D. melanogaster* larvae of oxygen for relatively short periods of time (up to 4 h) does not result in any changes in puffing. As soon as the larvae are restored to aerobic conditions, however, puffs rapidly appear *de novo* at nine loci — in fact the same nine puffs that are induced by temperature shock. Induction of these puffs occurs within five minutes after the end of anaerobic conditions and the size of the induced puffs, and the length of time they remain active, is dependent upon the duration of the previous anaerobic treatment. Later during the recovery period additional puffs are induced in *D. melanogaster* at several loci (X: 10EF, 2L: 26A, 2R: 45E, 48B, 53BC, 55EF, 3L: 76D, 3R: 84E, 85B, 86F, 96A and 100EF). The regularity of induction of these puffs, and their size, is less than that found for the initial nine puffs. (ASHBURNER, 1970c).

D. Miscellaneous Agents

RITOSSA (1964) reported that various chemical uncouplers of oxidative phosphorylation (3×10^{-3}M sodium azide, 10^{-2}M sodium salicylate, 10^{-3}M dicumarol, 10^{-3}M dinitrophenol) would induce the temperature shock series of puffs in isolated salivary glands. Higher (10^{-2}M) 2, 4, dinitrophenol, did not induce puffs; neither did the metabolic inhibitors iodoacetate (10^{-3}M) or sodium fluoride (10^{-2}M).

In vitro incubation of isolated salivary glands in various media alone results in specific puff induction within short periods of time (less than 30 minutes). The following puffs have been observed in several strains of *D. melanogaster* as the result of salivary gland incubation in a variety of media [simple Ringer type solutions such as BEADLE and EPHRUSSI Ringer or BECKER's (1959) Ringer, buffered saline media or complex media (e.g. Grace's TC medium)]: X: 5C, 2R: 50CD, 50F, 57D, 3R: 84E, 85B (ASHBURNER, unpublished; DAVIS, unpublished). Indeed even the injection into L3 of as little as 0.08 µl of 10% ethanol in water will result in the induction or enlargement of these puffs within 15 min.

No systematic study of the effects of media differing in their ionic composition on puffing in *D. melanogaster* have been carried out although BERENDES et al. (1965) have done some work along these lines with *D. hydei*.

Incubation, *in vitro*, of salivary glands in media containing antibiotic inhibitors of protein synthesis (cycloheximide, puromycin) is without effect on puffing in *D. melanogaster* (ASHBURNER, 1970c; ELLGAARD, unpublished). Yet injection of cycloheximide or puromycin into 90 h L3 has marked effects on puffing. The treatment inhibits the incorporation of amino acids into proteins almost immediately (less than 1 minute). The changes in puffing, however, take about 3 h to be apparent. Most obvious is the induction of two puffs, at 2L: 23C and 3L: 73A, sites not normally seen to be active. Smaller puffs are also induced at 3R: 91B, 96E, 3L: 66B, 74EF and 75B. These five sites are normally active during development (ASHBURNER and MITCHELL, unpublished, see also the papers of CLEVER 1966; 1967; SERFLING, PANITZ and WOBUS, 1969).

Table 7. *Miscellaneous agents and their reported effects on puffing in* D. melanogaster

Agent	Effect	Notes	References
1. Tryptophan 0.03 M	Specific induction *in vitro* of puff 68D	Not repeatable by MILKMAN (personal communication) or ASHBURNER (1970a)	FEDEROFF and MILKMAN (1964, 1965)
2. Dicyandiamide 0.1 M and 0.01 M	Increase in puffing activity at various sites (*in vitro*)	Not controlled by sister glands or for larval age	MUKHERJEE (1968)
3. Cortisone acetate	No effects on puffing when fed to larvae at 800 mg% on axenic medium (SANG), or on yeast medium (SMITH et al.), or *in vitro* (up to 1 mg/ml)	Contra GOODMAN et al. (1967)	SANG (1968), SMITH, KOENIG and LUCCHESI (1968)
4. Hydrocortisone phosphate	Induction of developmental specific puffs	Not controlled by sister glands or for larval age	GILBERT and PISTEY (1966)
5. EDTA	Heritable supernumary nucleoli after larval feeding	—	KHRISTOLYUBOVA (1961) KHRISTOLYUBOVA and AUSLANDER (1967)
6. Histones (lysine rich) and polylysine	Coinjection of 2 μg (?) with 15 CU ecdysone inhibits ecdysone induction	Not controlled by injection of non histone basic proteins	DESAI and TENCER (1968)
7. Drosophila embryo extracts	Specific induction in glands incubated *in vitro* in medium plus egg or L1 extracts	Not controlled	TOKUMITSU (1968)
8. Roux Sarcoma Virus	Reduction in size of puffs after feeding to larvae	—	BURDETTE and YOON (1967)

A number of other compounds have been tested for their effect on puffing in *Drosophila melanogaster*. Unfortunately almost without exception these studies have been inadequately controlled and have failed to take into account the developmental variation in puffing activity. For these reasons they cannot be fruitfully discussed and are merely listed in Table 7.

VII. Other Studies on Puffing in *D. melanogaster*

A. Dosage Compensation

Autosomal puffing patterns of *D. melanogaster* and *D. hydei* appear identical in male and female larvae (BERENDES, 1965; ASHBURNER, 1967). In *D. melanogaster* and *D. simulans* there are quantitative differences in puffing at some X chromosome loci: in both species the puffs at 2B 5—6, 2B 13—17 and 2EF are active longer following puparium formation in male than in female larvae (ASHBURNER, 1969a). Sex differences at a few X chromosome loci have also been detected in *D. ananassae* (GUPTA, 1968).

Autoradiographic studies of RNA synthesis in *D. melanogaster* by MUKHERJEE and BEERMANN (1965), MUKHERJEE (1966), MUKHERJEE, LAKHOTIA and CHATTERJEE (1968) KAPLAN and PLAUT (1968) and by KORGE (1970) have attempted to analyse the mechanisms of dosage compensation for X chromosome genes. These experiments suggest that hyperactivity of the single male X chromosome is, at least in salivary glands, the basis of dosage compensation.

The "haploid" X chromosome of male *Drosophila* is, in polytene chromosome preparations, more diffuse in appearance than the "diploid" X chromosomes of females [OFFERMANN (1936), see DOBZHANSKY (1957)]. The "haploid" homologue of an asynapsed X chromosome in females and the "haploid" regions in females heterozygous for deletions are never of a similar morphology to the male X. The diameter of a male X chromosome is greater than that of a 'haploid' asynapsed female homologue: the ratios of the diameters of synapsed female XX: asynapsed female X: male X are given as 1.00: 0.68: 0.91 for *D. melanogaster* by MUKHERJEE, LAKHOTIA and CHATTERJEE (1968) and 1.00: 0.58: 0.83 for *D. hydei* by BERENDES (1966). X-ray irradiation of larvae temporarily reduces the diameter of the male X chromosome, while having little effect on that of female X chromosome (MUKHERJEE, LAKHOTIA and CHATTERJEE (1968): see discussion in MULLER and KAPLAN (1966).

The DNA content of the male polytene X chromosome is half that of the female X (ARONSON, RUDKIN and SCHULTZ, 1954; RUDKIN, 1965). The protein content of the male X is, however, disproportionately high (RUDKIN, 1965).

Incorporation of tritiated uridine by male and female X chromosomes of *D. melanogaster* was measured by MUKHERJEE and BEERMANN (1965) and by MUKHERJEE (1966). The incorporation of label into the X chromosomes was compared with the incorporation into a standard autosomal region. Four comparisons were made: X (tip to 3B) *vs.* 3L: 61, X (tip to 3B) *vs.* 3L: (tip to 68B), whole X *vs.* 3L: 61 and whole X *vs.* 3L (tip to 68B) in males and females. The grain count ratios in each case were identical in male and female larvae. For a standard X chromosome region the number of grains was higher in the male X than in an asynapsed female X chromosome homologue. In a very original study LAKHOTIA and MUKHERJEE (1969) show that the hyperactivity of the male X is cell autonomous. Somatic nondisjunction of an unstable ring X chromosome in female heterozygous for In (X^{c2}) w^{rc} and a rod chromosome results in mosaic salivary glands with genotypically male (XO) nuclei together with XX female nuclei The morphological appearance of the X chromosome in the XO nuclei is similar to its appearance in normal males. Likewise the ratio of grains resulting from H^3U incorporation into the X of "male XO nuclei" and the XX of adjacent "female nuclei" is similar to that found in XY male and XX female larvae.

Korge (1970) confirmed Mukherjee's results. He also analysed the effects on uridine incorporation into X chromosomes of various duplications and deficiencies. In structurally normal X chromosomes of male or female white prepupae sacrificed 15 min after isotope injection between 18 and 21% of the silver grains of X (tip to 3B) lie over the puff in region 3B (3A of Ashburner, 1969a). Female larvae heterozygous for $Df(1)w^{258-11}$ are heterozygous for a deficiency of the 3B puff. In these heterozygotes there is dosage compensation for the deficiency, the hemizygous 3B puff incorporates approximately 16% of the label of the X tip-3B region, compared with 18% in wild type females.

There was no dosage compensation in the two duplications studied, $Dp(1:f)$ 101 and $Dp(1:3)$ w^{rco}. The duplicated chromosome regions having a higher grain count than in wild type chromosomes. This 'dosage effect' was especially pronounced in duplication carrying males.

In a brief abstract Kaplan and Plaut (1968) also confirm the data of Mukherjee and Beermann with respect to H^3 uridine incorporation into male and female X chromosomes. They report a lack of any Y chromosome effect on uridine incorporation into the X chromosome in males as judged by a comparison of labelling in XY and XO larvae.

B. Circadian Rhythms

Many physiological processes show diurnal (circadian) variation in their activities. Rensing (1966) found that the nuclear volume of the *Drosophila* prothoracic gland changes diurnally and later Rensing and Hardeland (1967) investigated puffing patterns in animals grown under controlled light: dark cycles. In L3 the maximum mean PS (PS 6) was reached 3—6 h into the light period when the larvae were grown under a 12 : 12 h day at 25° C. Under these conditions there is variation in the frequency of puparium formation, a peak occurring 9 h into the light phase. The peak in mean puff stage preceded this peak in puparium formation by 6—3 h; an observation quite consistent with that on animals grown in uncontrolled light conditions (see section II).

C. Salivary Gland Transplantation Experiments

Becker (1962a, b) transplanted salivary glands of *D. melanogaster* between larvae or prepupae of differing ages. After a short period of culture *in vivo* the puffing patterns of the transplant were compared with those of the host. The general, and important, conclusion of this study was that the puffing pattern of the transplant was similar to that of the host, rather than similar to the donor pattern. Glands from PS 1 L3 were transplanted into older L3 and there was premature induction of their puffs. Salivary glands from PS 13 (2 h) prepupae were transplanted into larvae and there was a re-induction of a characteristic larval puffing pattern on 3L (puffs 62E, 74EF, 75B and 78D). In many transplants the PS of the implant was slightly delayed with respect to the PS of the host. This may be a result of the unusual posterior location of the transplanted salivary gland.

In STAUB's (1969) experiments salivary glands of L3 and prepupae were transplanted and cultured in the abdomens of adult hosts [see BERENDES and HOLT, (1965) who did similar work in *D. hydei*]. Abnormal puff behaviour was found in chromosome arm 3L following *in vivo* culture of salivary glands in this environment. The actual puffs induced as the result of this treatment varied and depended on two main factors: (a) age of donor animal; (b) age and physiological condition

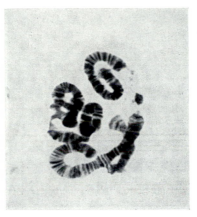

Fig. 33. "Super giant" salivary gland chromosomes from a 7 h embryo of *D. melanogaster* after culture in an adult abdomen for 13 days. (From STAUB, 1969)

(male *vs.* female, virgin *vs.* nonvirgin female) of the host. Two abnormal puffs at 68B and 78E on 3L were induced in PS 1 glands transplanted into female hosts. At 78E puffing was first observed 6 h after implantation and was most frequent in glands transplanted to old (10—15 days) virgin female hosts. The puff 68B was more frequent when the hosts were either very young females (within four hours of eclosion) or older males.

Transplantation of prepupal salivary glands led to induction of a third puff, at 63BC (see p. 121 and 139) and also of the previously noted puffs at 68B and 78E.

The physiological variation in host conditions resulting in these differences in puffing of implanted glands are not understood. To some extent the chromosomal reaction probably reflects the different hormonal environments of the adult hosts.

In further experiments STAUB transplanted into adult hosts fragments of embryos of *D. melanogaster*. The fragments, front halves of embryos, included the salivary gland rudiments and during prolonged culture (5—13 days) *in vivo* these cells differentiated and developed polytene chromosomes. In some cells the chromosomes were similar to the "super giant" chromosomes found by HADORN, GEHRING and STAUB (1963) following prolonged culture in adult abdomens of L3 salivary glands. (Fig. 33). In STAUB's cultures puffs on 3L were observed at both developmentally normal loci (62F, 71DE) and at 68B and 78E.

VIII. Conclusion

Having summarized much of what is known of the behaviour of polytene chromosome puffs in *D. melanogaster* it is clear that our knowledge is far from complete. The great potential for studies with this species stems from the relative ease of its genetic analysis. Although some puffs are under a fairly direct hormonal control we are far from understanding the genetic organization of the control systems. Due to a lack of puff: phenotype correlations we are even further from understanding the relationship of these control systems to the biochemical economy of the tissue and of the organism. Without doubt answers to these problems will depend upon parallel physiological, biochemical and genetic studies of puffing at all its levels of analysis. *Drosophila melanogaster* is an organism well suited for this task.

Acknowledgements

I am grateful to those persons who have allowed me to quote their unpublished observations and to republish their figures. I wish to thank Professor J. M. Thoday, and Dr. S. A. Henderson for their help with the manuscript of this review and Mrs. C. Finnegan for assistance with the illustrations.

Appendix

A list of puff loci in *D. melanogaster* with their precise positions (when known) according to the revised maps of Bridges (see Lindsley and Grell, 1968). In the text and figures abbreviated positions are cited (e.g. 63E for 63E1—6) unless this would result in ambiguity. Synonyms — locations cited in previous papers and now revised — are given in parenthesis.

Chromosome 1	14B	28A1—2(28C)	46D
1C	15A	28D1—2	46F5—6
2B5—6	15C3—6	29BC	47A9—16
2B13—17	16A4—6	29F1—2	47BC
2EF	16C	30A3—6	47F
3A1—4	16DE	30B	48A
3AB	18D	30E	48B
3C11—12		31A	48E
3E	Chromosome 2	32C5—D1/2	49B
4F1—4	21B	33B	49D
4F9—10	21C4—6	33E3—8	49E(49F)
5B	21D	34A5—6	50A3—4
6B	21F	35B1—3(35A)	50C6—10
6F	22A	36F6?—7?	50C23—D4
7D14—15	22B4—5	37D1—3	50F
8D	22B8—9	38B1—2(37F)	51D
8EF	22C3/4—6	39BC(38A)	52A
9A	23C (23B)	42A	52C4—7
9EF	23E	42B	53CD
10EF	24E	43B	54BC
11B14—17	25AC	43E	54E(54F)
11E	25D	44AB	55B
12A4—5	26A	44F1—2	55E
12E3—7	26B10—11(26AB)	45E	56D
13E1—2	26C	45F1—2	57C
13E7—16	27C1—2	46A1—2	57D8—9

57E	64E(64F)	77B5—7(77AB)	90BC
58A1—3	65B	77E	91D1—6(91B)
58BC	66B	78D1—5	92A
58DE	66E1—2	78E	93B
59D	67B	79D	93D1—2
59F4—8	67F	82CD	93F9—10
60A	68B	82F8—9(82EF)	94B
60B7—13	68C	83AB	94EF
60D	69A1—3	83E	95B
60E	70A	84F1—2(84E)	95D
	70C	85B	95F
Chromosome 3	70E	85D1—2	96A
61C	71B(71AB)	85F1—6	96E
62A	71DE(71CE)	85F10—16	97C
62B	72D(72CD)	86E	98B
62C	73A1—2	86F	98D
62E	73B5—7	87A	98F1—2
62F	73F	87B	99B
63BC	74C	87C	99D
63E1—3	74EF	87D	99E
63F	75B	87F	100B
64A11—13	75CD	88D	100DE
64B10—14	76A(76B)	88EF	Chromosome 4
64C9—13	76D1—4	89B9—22	102C10—17(102CD)

References

ARONSON, J. F., RUDKIN, G. T., SCHULTZ, J.: A comparison of giant X-chromosomes in male and female *Drosophila melanogaster* by cytophotometry in the ultraviolet. J. Histochem. Cytochem. **2**, 458—459 (1954).

ASHBURNER, M.: Patterns of puffing activity in the salivary gland chromosomes of *Drosophila*. I. Autosomal puffing patterns in a laboratory stock of *D. melanogaster*. Chromosoma (Berl.) **21**, 398—428 (1967a).

— Gene activity dependent on chromosome synapsis in polytene chromosomes of *Drosophila melanogaster*. Nature (Lond.) **214**, 1159—1160 (1967b).

— Patterns of puffing activity in the salivary gland chromosomes of *Drosophila*. II. X chromosome puffing patterns in *D. melanogaster* and *D. simulans*. Chromosoma (Berl.) **27**, 47—63 (1969a).

— Patterns of puffing activity in the salivary gland chromosomes of *Drosophila*. III. A comparison of the autosomal puffing patterns of the sibling species *D. melanogaster* and *D. simulans*. Chromosoma (Berl.) **27**, 64—85 (1969b).

— Patterns of puffing activity in the salivary gland chromosomes of *Drosophila*. IV. Variability of puffing patterns. Chromosoma (Berl.) **27**, 156—177 (1969c).

— The genetic control of puffing in polytene chromosomes. In: DARLINGTON, C.D., LEWIS, K.R. (Eds.): Chromosomes Today, Vol. 2, pp. 99—106. Edinburgh: Oliver and Boyd 1969d.

— On the problem of the genetic similarity between sibling species; Puffing patterns in *Drosophila melanogaster* and *Drosophila simulans*. Amer. Naturalist **103**, 189—191 (1969e).

— Function and structure of polytene chromosomes during insect development. Advan. Insect Physiol. **7**, 1—95 (1970a).

— Effects of juvenile hormone on adult differentiation of *Drosophila melanogaster*. Nature (Lond.) **227**, 187—189 (1970b).

— Patterns of puffing activity in the salivary gland chromosomes of *Drosophila*. V. Responses to environmental treatments. Chromosoma (Berl.) **31**, 356—376 (1970c).

— The Genetic analysis of puffing in polytene chromosome of *Drosophila*. Proc. Roy. Soc. B. **176**, 319—327 (1970d).

Baker, W. K.: Position-effect variegation. Advan. Genetics **14**, 133—169 (1968).

Becker, H. J.: Die Puffs der Speicheldrüsenchromosomen von *Drosophila melanogaster*. I. Beobachtungen zum Verhalten des Puffmusters im Normalstamm und bei zwei Mutanten giant und lethal-giant-larvae. Chromosoma (Berl.) **10**, 654—678 (1959).

— Die Puffs der Speicheldrüsenchromosomen von *Drosophila melanogaster*. II. Die Auslösung der Puffbildung, ihre Spezifität und ihre Beziehung zur Funktion der Ringdrüse. Chromosoma (Berl.) **13**, 341—384 (1962a).

— Stadienspezifische Genaktivierung in Speicheldrüsen nach Transplantation bei *Drosophila melanogaster*. Zool. Anz. Suppl. **25**, 92—101 (1962b).

Berendes, H. D.: Salivary gland function and chromosomal puffing patterns in *Drosophila hydei*. Chromosoma (Berl.) **17**, 35—77 (1965).

— Differential replication of male and female X-chromosomes in *Drosophila*. Chromosoma (Berl.) **20**, 32—43 (1966).

— The hormone ecdysone as effector of specific changes in the pattern of gene activities of *Drosophila hydei*. Chromosoma (Berl.) **22**, 274—293 (1967).

— Holt, Th. K. H.: The induction of chromosomal activities by temperature shocks. Genen en Phaenen. **9**, 1—7 (1964).

— Breugel, F. M. A. van, Holt, Th. K. H.: Experimental puffs in salivary gland chromosomes of *Drosophila hydei*. Chromosoma (Berl.) **16**, 35—46 (1965).

Breugel, F. M. A. van, Bos, H. J.: Some notes on differential chromosome activity in *Drosophila*. Genetica **40**, 359—378 (1969).

Bridges, C. B.: Salivary chromosome maps. J. Heredity **26**, 60—64 (1935).

Burdette, W. J., Yoon, J. S.: Mutations, chromosomal aberrations, and tumors in insects treated with oncogenic virus. Science **155**, 340—341 (1967).

Burnet, B., Hartmann-Goldstein, I. J.: The effects of ribonucleic acid deprivation on puffing in the salivary gland chromosomes of *Drosophila melanogaster*. Genet. Res. (Cambridge) **17**, 113—124 (1971).

Chaudhuri, A. R.: Lipoprotein nature of the so-called vacuole in salivary gland cells of the mutant "fat" in *Drosophila melanogaster*. Drosophila Information Service **44**, 118 (1969).

Clever, U.: Genaktivitäten in den Riesenchromosomen von *Chironomus tentans* und ihre Beziehungen zur Entwicklung. I. Genaktivierungen durch Ecdyson. Chromosoma (Berl.) **12**, 607—675 (1961).

— Induction and repression of a puff in *Chironomus tentans*. Devel. Biology **14**, 421—438 (1966).

— Control of chromosome puffing. In: Goldstein, L. (Ed.): The Control of Nuclear Activity. Englewood Cliffs, N. J.: Prentice Hall 1967.

Desai, L., Tencer, R.: Effects of histones and polylysine on the synthetic activity of the giant chromosomes of salivary glands of Dipteran larvae. Expt. Cell Res. **52**, 185—197 (1968).

Dobzhansky, T.: The X-chromosome in the larval salivary glands of hybrids *Drosophila insularis* × *D. tropicalis*. Chromosoma (Berl.) **8**, 691—698 (1957).

Ellgaard, E. G., Brosseau, G. E.: Puff forming ability as a function of chromosomal position in *Drosophila melanogaster*. Genetics **62**, 337—341 (1969).

Federoff, N., Milkman, R. R.: Specific puff induction by tryptophan in *Drosophila* salivary chromosomes. Biol. Bull. (Woods Hole) **127**, 369 (1964).

— — Induction of puffs in *Drosophila* salivary chromosomes by amino acids. Drosophila Information Service **40**, 48 (1965).

Fraenkel, G.: A hormone causing puparium formation in the blowfly *Calliphora erythrocephala*. Proc. roy. Soc. B **118**, 1—12 (1935).

Freire-Maia, N., Freire-Maia, A.: O Fenomeno de Pavan-Breuer no Genero Drosophila. V. Reuniao Anual da S. B. P. C., Curitiba (1953).

Gilbert, E. F., Pistey, W. R.: Chromosomal puffs in *Drosophila* induced by Hydrocortisone phosphate. Proc. Soc. exp. Biol. (N. Y.) **121**, 831—832 (1966).

Goodman, R. M., Goidl, J., Richart, R. M.: Larval development in *Sciara coprophila* without the formation of chromosomal puffs. Proc. nat. Acad. Sci. (Wash.) **58**, 553—559 (1967).

GROB, H.: Entwicklungsphysiologische Untersuchungen an den Speicheldrüsen, dem Darmtraktus und den Imaginalscheiben einer Letalrasse (*lgl*) von *Drosophila melanogaster*. Z. indukt. Abstamm.- u. Vererb.-Lehre. **84**, 320—360 (1952).

GUPTA, A. K. D.: Certain aspects of structural and functional differentiation of heterochromatin in salivary gland chromosomes of *Drosophila ananassae*. Seminar on Chromosomes. The Nucleus, Suppl. 340—344 (1968).

HADORN, E.: An accelerating effect of normal 'ring glands' on puparium-formation in lethal larvae of *Drosophila melanogaster*. Proc. nat. Acad. Sci. (Wash.) **23**, 478—484 (1937).

— GEHRING, W., STAUB, M.: Extensives Größenwachstum larvaler Speicheldrüsenchromosomen von *Drosophila melanogaster* im Adultmilieu. Experienta (Basel) **19**, 530—531 (1963).

HARTMANN-GOLDSTEIN, I.: On the relationship between heterochromatinization and variegation in Drosophila, with reference to temperature sensitive periods. Genet. Res. (Cambridge) **10**, 143—159 (1967).

HORTON, I. H.: A comparison of the salivary gland chromosomes of *Drosophila melanogaster* and *D. simulans*. Genetics **24**, 243—234 (1939).

KAPLAN, R. A., PLAUT, W.: A radioautographic study of dosage compensation in *Drosophila melanogaster*. J. Cell Biol. **39**, 71 a (1968).

KARLSON, P., HANSER, G.: Über die Wirkung des Puparisierungshormons bei der Wildforms und der Mutante *lgl* von *Drosophila melanogaster*. Z. Naturforsch. **7** b, 80—83 (1952).

KERKIS, J. J.: Chromosomal conjugation in hybrids between *Drosophila melanogaster* and *D. simulans*. Amer. Naturalist **70**, 81—86 (1936).

— The causes of imperfect conjugation of chromosomes in hybrids of *Drosophila melanogaster* and *simulans*. Izv. Akad. Nauk. SSR 459—468 (1937).

KHRISTOLIUBOVA, N. B.: Controlled alterations of the physiological activity of certain sections of giant chromosomes of the salivary glands of *Drosophila* as the result of the action of versene. Dokl. Akad. Nauk. SSR **138**, 681—682 (English Transl. Ed.) **138**, 382—384 (1961).

— AUSLÄNDER, I. E.: Unique inheritance of functional chromosome changes in salivary gland cells of *Drosophila melanogaster*. Genetika (USSR) **3**, 76—79 (In Russian with English summary). (1967).

KIKNADZE, I. I.: The structural and cytochemical characteristics of chromosome puffs. Symposium on the Mutational Process: Genetic Variations in Somatic Cells. 177—181 Prague. (1966).

KING, R. C., SANG, J. H., LETH, C. B.: The hereditary ovarian tumours of the *fes* mutant of *Drosophila melanogaster*. Expt. Cell Res. **23**, 108—117 (1961).

KOBEL, H. R., VAN BREUGEL, F. M. A.: Observations on *ltl* (lethal tumourous larvae) of *Drosophila melanogaster*. Genetica **38**, 305—327 (1961).

KORGE, G.: Dosage compensation and effect for RNA synthesis in chromosome puffs of *Drosophila melanogaster*. Nature (Lond.) **225**, 386—388 (1970).

LAKHOTIA, S. C., MUKHERJEE, A. S.: Chromosomal basis of dosage compensation in *Drosophila*. I. Cellular autonomy of hyperactivity of the male X-chromosomes in salivary glands and sex differentiation. Genet. Res. (Cambs) **14**, 137—150 (1969).

LAMPRECHT, J., REMENSBERGER, P.: Polytäne Chromosomen im Bereiche der Prothoracaldorsal-Imaginalscheibe von *Drosophila melanogaster*. Experientia (Basel) **22**, 293 (1966).

LAUFER, H., GREENWOOD, H.: The effects of juvenile hormone on larvae of the dipteran, *Chironomus thummi*. Amer. Zool. **9**, 603 (1969).

— HOLT, T. H. K.: Juvenile hormone effects on chromosomal puffing and development in *Chironomus thummi*. J. exp. Zool. **173**, 341—351 (1970).

LEWIS, E. B.: The phenomenon of position effect. Adv. Genetics **3**, 73—115 (1950).

LEZZI, M., GILBERT, L. I.: Control of gene activities in the polytene chromosomes of *Chironomus tentans* by ecdysone and juvenile hormone. Proc. nat. Acad. Sci. (Wash.) **64**, 498—503 (1969).

LINDSLEY, D. L., GRELL, E. M.: Genetic variations of *Drosophila melanogaster*. Carnegie Inst. Wash. Publ. **627**, pl-472 (1968).

MORIWAKI, D., ITO, S.: Studies on puffing in the salivary gland chromosomes of *Drosophila ananassae*. Jap. J. Genetics **44**, 129—138 (1969).

Mukherjee, A. S.: Dosage compensation in *Drosophila*: An autoradiographic study. Nucleus **9**, 83—96 (1966).
— Effects of dicyandiamide on puffing activity and morphology of salivary gland chromosomes of *Drosophila melanogaster*. Ind. J. exp. Biol. **6**, 49—51 (1968).
— Beermann, W.: Synthesis of ribonucleic acid by the X-chromosomes of *Drosophila melanogaster* and the problem of dosage compensation. Nature (Lond.) **207**, 785—786 (1965).
— Lakhotia, S. C., Chatterjee, S.: On the molecular and chromosomal basis of dosage compensation in *Drosophila*. Seminar on Chromosomes. Nucleus Suppl. 161—173 (1968).
Muller, H. J., Kaplan, W. D.: The dosage compensation of *Drosophila* and mammals as showing the accuracy of the normal wild type. Genet. Res. (Cambs) **8**, 41—59 (1966).
Offermann, C. A.: Branched chromosomes as symmetrical duplications. J. Genet. **32**, 103—116 (1936).
Pätau, K.: Chromosomenmorphologie bei *Drosophila melanogaster* und *Drosophila simulans* und ihre genetische Bedeutung. Naturwissenschaften **23**, 537—543 (1935).
Pätterson, J. T., Stone, W. S.: Evolution in the Genus *Drosophila*, 610 p. New York: MacMillan 1952.
Rayle, R. E.: A new mutant in *Drosophila melanogaster* causing an ecdysone-curable interruption of the prepupal molt. Genetics **56**, 583 (1967).
— Hoar, D. I.: Gene order and cytological localisation of several X-linked mutants of *Drosophila melanogaster*. Drosophila Information Service **44**, 94 (1969).
Rensing, L.: Zur circadianen Rhythmik des Hormonsystems von *Drosophila*. Z. Zellforsch. **74**, 539—538 (1966).
— Hardeland, R.: Zur Wirkung der circadianen Rhythmik auf die Entwicklung von *Drosophila*. J. Insect Physiol. **13**, 1547—1568 (1967).
Ritossa, F.: A new puffing pattern induced by temperature shock and DNP in *Drosophila*. Experientia (Basel) **18**, 571—573 (1962).
— New puffs induced by temperature shock, DNP and salicilate in salivary chromosomes of *D. melanogaster*. Drosophila Information Service **37**, 122—123 (1963).
— Experimental activation of specific loci in polytene chromosomes of *Drosophila*. Exp. Cell Res. **35**, 601—607 (1964).
Roberts, P. A.: Position effect on DNA transcription and replication. Genetics **61**, s 50 (1969).
— Behaviour of a position effect puff in *Drosophila melanogaster*. Genetics **64**, s 53 (1970).
Rodman, T.: The larval characteristics and salivary gland chromosomes of a tumorigenic strain of *Drosophila melanogaster*. J. Morph. **115**, 419—446 (1964).
— Kopac, M. J.: Alterations in morphology of polytene chromosomes. Nature (Lond.) **202**, 876—877 (1964).
Rudkin, G. T.: The ultraviolet absorption of puffed and unpuffed homologous regions in the salivary gland chromosomes of *Drosophila melanogaster*. Genetics **40**, 593 (1955).
— The relative mutabilities of DNA in regions of the X-chromosome of *Drosophila melanogaster*. Genetics **52**, 665—681 (1965).
— Woods, P. S.: Incorporation of H^3 cytidine and H^3 thymidine into giant chromosomes of *Drosophila melanogaster* during puff formation. Proc. nat. Acad. Sci. (Wash.) **45**, 997—1003 (1959).
Russell, E. S.: A comparison of benign and 'malignant' tumours in *Drosophila melanogaster*. J. exp. Zool. **84**, 363—386 (1940).
Sang, J. H.: Lack of cortisone inhibition of chromosomal puffing in *Drosophila melanogaster*. Experientia (Basel) **24**, 1064 (1968).
Scharrer, B., Hadorn, E.: The structure of the ring gland (corpus allatum) in normal and lethal larvae of *Drsoophila melanogaster*. Proc. nat. Acad. Sci. (Wash.) **24**, 236—242 (1938).
Schultz, J.: Genes, Differentiation, and Animal Development. Brookhaven Symposium No. 18, 116—147 (1965).
Serfling, E., Panitz, R., Wobus, U.: Die experimentelle Beeinflussung des Puffmusters von Riesenchromosomen. I. Puffinduktion durch Oxytetracyclin bei *Chironomus thummi*. Chromosoma (Berl.) **28**, 107—119 (1969).

SLIZYNSKI, B.M.: Functional changes in polytene chromosomes of *Drosophila melanogaster*. Cytologia (Tokyo) **29**, 330—336 (1964).

SMITH, P.D., KOENIG, P.B., LUCCHESI, J.C.: Inhibition of development in *Drosophila* by cortisone. Nature (Lond.) **217**, 1286 (1968).

STAUB, M.: Veränderungen im Puffmuster und das Wachstum der Riesenchromosomen in Speicheldrüsen von *Drosophila melanogaster* aus spätlarvalen und embryonalen Spendern nach Kultur *in vivo*. Chromosoma (Berl.) **26**, 76—104 (1969).

TOKUMITSU, T.: Some aspects on effects of *Drosophila* tissue extracts on the puffing pattern of incubated *Drosophila* salivary glands. J. Fac. Sci. Hokkaido Univ. Ser. Zool. **16**, 525—530 (1968).

WELCH, R.M.: A developmental analysis of the lethal mutant 1(2)gl of *Drosophila melanogaster* based on cytophotometric determination of nuclear desoxyribonucleic acid (DNA) content. Genetics **42**, 544—559 (1957).

Relation of Puffing to Bristle and Footpad Differentiation in Calliphora and Sarcophaga[1]

DIETER RIBBERT

Zoologisches Institut der Universität Münster, Münster

I. Introduction

Differentiation is the central problem during the organization of a multicellular organism. Sufficient experimental evidence has been collected to show that differential cell functions result from differential gene activation and repression.

Most of our knowledge about the storage, reduplication and regulated expression of information has been derived from experimental results obtained in detailed studies of prokaryotes; but we must expect the processes of information transfer in eukaryotes to be much more complicated, though the basic mechanisms may be the same in pro- and eukaryotes. Apart from methodical limitations in the investigations of eukaryotes, it is probable that biochemical pathways of gene expression in this category of organisms are far more complex and this may be the main reason why the elucidation of these processes is proceeding less rapidly than in the case of prokaryotes.

Among the most favorable systems for analyzing these problems by cytogenetic and biochemical methods are the polytene chromosomes. The puffing phenomenon of the polytene chromosomes has been interpreted as the morphological pendant of differential gene activity (BEERMANN, 1952, 1956, 1959, 1965). This idea has been supported by a considerable amount of experimental data showing that the puffing of the chromosomes follows a tissue and phase specific pattern (BEERMANN, 1952; MECHELKE, 1953; BECKER, 1959; CLEVER and KARLSON, 1960; CLEVER, 1961; PANITZ, 1960, 1964; BERENDES, 1965; ASHBURNER, 1967) and by the discovery that the puffs are those loci of the chromosomes where a rapidly labeled RNA is synthesized (PELLING, 1959, 1964).

The real problem, however, is to establish a direct correlation between the synthesis of specific proteins (as manifestation of the differentiated state of a cell) and the tissue-specific puffs, which should represent the activated genes controlling the synthesis of the proteins. There is some tentative experimental evidence in support of this correlation. In *Acricotopus lucidus* the simultaneous presence of two Balbiani rings and specific secretions in distinct parts of the salivary gland has been demonstrated (BAUDISCH, 1960, 1961; MECHELKE, 1953, 1958). A direct relationship between one of the Balbiani rings in question and the ability

---

1 The investigations of Calliphora trichogen cells have been supported by the Deutsche Forschungsgemeinschaft.

of the cell to secrete saliva containing hydroxyproline has been made plausible by the demonstration of the fact that, under the influence of Gibberelline A3, RNA synthesis of this Balbiani ring is specifically reduced; at the same time the cells are no longer able to form hydroxyproline (Panitz, 1967; Baudisch and Panitz, 1968).

Two further examples could be cited to show the parallel occurrence of certain phenotypic features and specific puffs (Slyzinsky, 1964; Berendes, 1965), but only in the case of *Camptochironomus pallidivittatus* salivary glands has a direct correlation been established between the specific synthetic capacity of the so-called special cells and an additional Balbiani ring of these cells (Beermann, 1961). Grossbach (1969) has used the latter system to analyze the pattern of secretory proteins in different parts of the gland and he has shown the direct genetic relationship between the synthesis of specific protein fractions and distinct chromosome regions where tissue-specific Balbiani rings are located.

All these investigations have been undertaken with polytene chromosomes from salivary glands. But it is certain that the functional metabolism of salivary gland cells is only vaguely related to developmental events during metamorphosis. This is definitely an advantage for analysis such as has been effected in the case of the special cells of the salivary glands in *Camptochironomus pallidivittatus*. For the purpose of extending the potential field of experimentation it is nevertheless equally desirable to have a cell system which has easily analyzable polytene chromosomes, but which is more directly connected with development. In this respect the detection of large polytenic cells in the pupal footpads of *Sarcophaga bullata* (Whitten, 1964) and in the bristle-forming apparatus of various Dipteran species (Ribbert, 1967) may be just the experimental system we have been looking for.

In the following report these two closely related cell systems will be presented, and existing experimental data will be reviewed to demonstrate the value of these epidermal cells for future investigations of gene induction and regulation in euka-ryotes.

II. The Morphogenesis of Bristles and Footpads

A. The Bristle-forming Apparatus of Calliphora

The chromosomes of bristle-forming cells including their puffing pattern have so far been investigated in *Calliphora erythrocephala* (Ribbert, 1967) and recently in Sarcophaga barbata (Trepte, 1971). The bristle-forming apparatus in *Calliphora* directly involves only two cells, of which the larger one produces the long tapering hair (trichogen cell), while the smaller tormogen cell forms a circular chitinous socket around the basis of the bristle (cf. Fig. 1a, b). During the development in the pupa these two cells grow rapidly by endomitotic polyploidization, attaining a nuclear volume about three thousand times that of the neighboring diploid epidermal cells (cf. Fig. 1c).

Contrary to previous findings mainly based on observations of larval polytene chromosomes in Calyptratae, the chromosomes of the bristle-forming cells assume a polytenic banded form of surprisingly good quality (cf. Fig. 2). The severe fragmentation and unspecific fusing tendency characteristic of larval polytene chromosomes of Calyptratae are largely reduced, and it is possible to construct chromosome maps and to analyze the puffing pattern in relation to morphogenesis of the bristle.

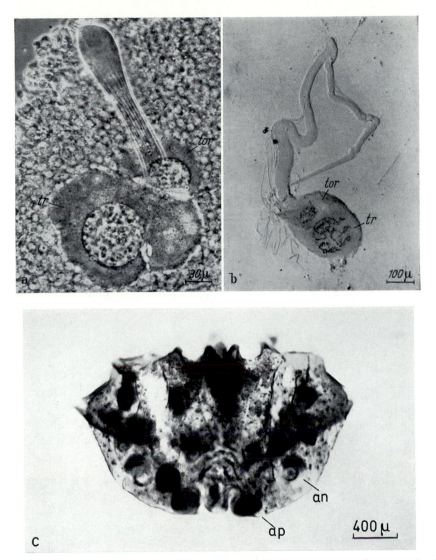

Fig. 1a—c. Bristles and bristle-forming cells. a unfixed preparation of a germinating bristle papilla at day 3.5 (phase contrast). In b a fully grown apical bristle at day 6 is seen. c isolated scutellum stained with aceto-carmine. The large bristle-forming cells with polytene nuclei are visible through the still unsclerotized, transparent cuticle. ap apical bristle-forming cell; an angular bristle-forming cell (RIBBERT, 1967)

The morphological analysis has been confined to the thoracic bristles and the cells which form them, since they can easily be prepared on account of their large size. Fig. 3 represents the arrangement of the macrochaetae on the thorax surface of *Calliphora erythrocephala*. A staging of the pupal development has been attributed to the formation of white prepupae. Under constant rearing conditions — especially a constant temperature of 21° C ± 0.5° C — a sample of white prepupae

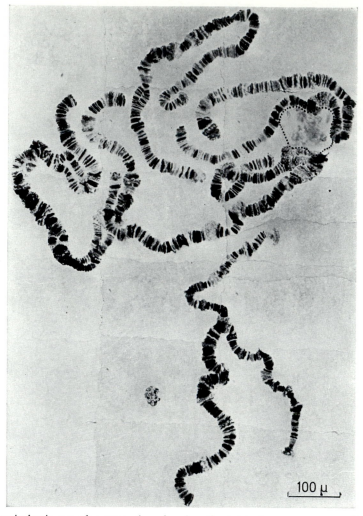

Fig. 2. Orcein-lactic squash preparation showing the total chromosome complement of a bristle-forming cell. The dotted line encircles the area of the nucleolus with adherent chromatin material of chromosome VI

of equal weight will develop very much more uniformly, and at the end of the puparium period (11 days) the adult flies of a batch will all emerge within four to six hours. According to the different volumes of the adult bristles, the trichogen cell nuclei (TCN) have also attained different final volumes. The accompanying tormogen cells are always smaller, with ratios between the two corresponding nucleus volumes differing between about 1/2, 1/4 and 1/8.

The change in the individual nuclear volume follows a characteristic scheme during the course of development (cf. Fig. 11); 2.5 days after pupation trichogen cells become conspicuous owing to the fact that the volume of their nuclei increases more markedly than in the neighboring diploid epidermal cells. The steep rise in the graph of Fig. 11 between days 3 and 5 represents the period of intensive endoreduplication, indicated by successive thickening of the chromosomes and by their ability to incorporate ^{3}H-thymidine. The highest nuclear volume in the trichogen cells is attained about the fifth day. This first period of rapid TCN growth is essentially due the successive polyploidization steps; the following changes in volume must reflect functional states of the nuclei with a general tendency towards gradual reduction from the fifth day onwards; but a second lower maximum of volume is apparent on the ninth

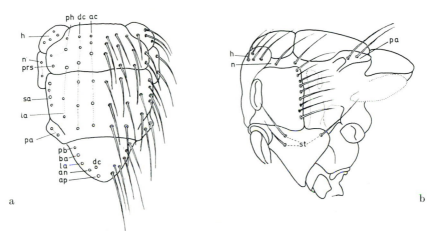

Fig. 3. Diagram of the arrangement of the large bristles on the thoracic surface of *Calliphora erythrocephala* with readily analyzable polytene chromosomes in their generating cells. a dorsal view, b profile view (RIBBERT, 1967)

day (cf. Fig. 11). If a growth factor of about 2 per endoreduplication step is taken into account, the largest TCN, 110 μ in diameter, which is that of the postalar bristle (pa in Fig. 3), would attain a maximum polyteny of 4096-n. For the majority of TCN in the mesothorax and scutellum (cf. Fig. 3) degrees of polyteny are probably between 1024-n and 2048-n at this time. The chromosomes of all these trichogen cells are of sufficient size and quality to permit an analysis of their puffing pattern from the fourth day onward at the latest.

The period of the strongest chromosomal endoreduplication coincides with the actual growth of the bristle. The germination of macrochaetal papillae becomes conspicuous on the third day after pupation (cf. Fig. 1). Young bristles, as well as certain areas of the cytoplasm close to the nucleus, exhibit a striking birefringence if observed between two crossed polaroids of a polarizing light microscope with four maxima of brightness per 360° rotation of the object. Similarly, LEES and PICKEN (1945) noted a birefringence in the initial processes of bristle growth of *Drosophila*. This indicates the formation of parallel structural elements in the bristle-forming cell. At the ultrastructural level it can be shown that at this time a cytoarchitectonic framework of densely packed microtubules is built up, which is obviously the constructive basis for the highly bipolar organization of the bristle-forming cell (cf. Figs. 4 and 5). Somewhat later, about the fifth day, in addition to the microtubules a fibrous cytoplasmic component, organized in columns, becomes visible in

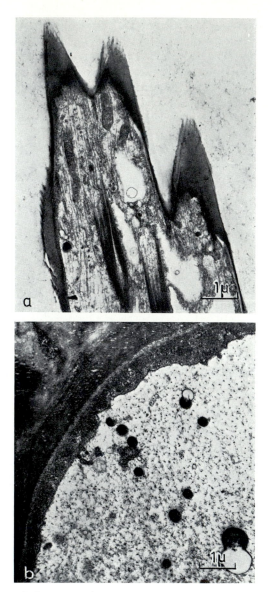

Fig. 4a and b. Microtubular cytosceleton. The precise parallel alignment of the microtubules can be seen in longitudinal sections a; in cross sections b there are many evenly
spaced microtubules, separated from each other by about 1200 Å. The electron-dense
vesicles may represent endocuticle precursors

the bristle shafts. This component is another element of the cytoskeleton or may
represent a chitinous (?) precursor to be incorporated into the cuticle (Fig. 6). In
principle, the cytoarchitectonic organization of the Calliphora bristle shafts resembles
that in other insects such as have been described in the case of bristles in *Oncopeltus*
(Lawrence, 1966), *Drosophila* bristles (Overton, 1967) or the hairs of a Lepidopter

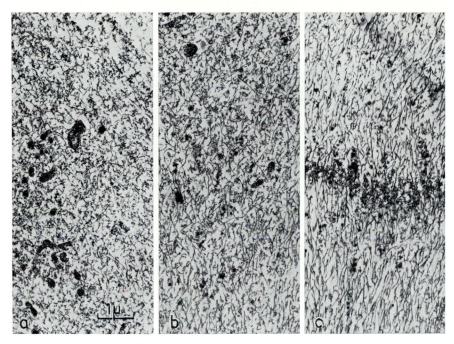

Fig. 5a—c. Microtubule arrangement with increasing parallel alignment in the transition from the nucleus-adjacent cytoplasmic areas a to the bristle shaft c (day 4)

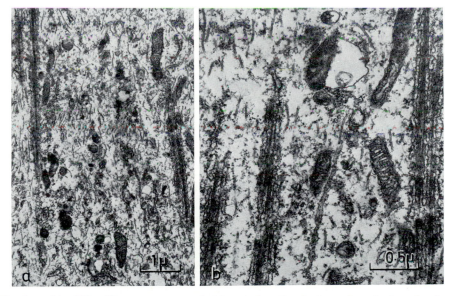

Fig. 6a and b. Microfilaments accompanied by tubules of endoplasmatic reticulum developing in a bristle at day 6

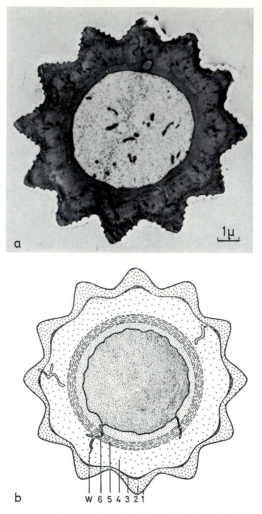

Fig. 7a and b. Cross sections through the tip of a bristle (late day 10) with a the secretion of endocuticle just beginning, b represents a somewhat later stage. 1—6 sequentially deposited cuticle layers; w wax canal

(Locke, 1969). Cross sections of the bristles show an evenly spaced arrangement of the microtubules, separated from one another by a centre-to-centre distance of about 1200 Å (Fig. 4b). The ultrastructure of the fibrous columns differs from similar structures described in other organisms in that the filamentous bundles are accompanied by regularly strung tubes of endoplasmatic reticulum (cf. Fig. 6.).

    At the end of the fourth day the bristles have completed growth and have attained their definitive length. It is at this time that deposition of cuticle substance begins at the surface of the bristle shafts. In the course of the following days until emergence, at least 6 different layers of cuticle, discernible by individual ultrastructure, will be secreted (cf. Fig. 7) in addition to the ecdysis membrane. This occurs in a strictly defined temporal sequence, the formation of the lamellated layer 5 having been completed late on the seventh day. Before the deposition of the last layer 6 (endocuticle), beginning on day 10.5, the effects of initial sclerotization on day 8 become visible. The cuticle of the bristles now exhibits an amber colour. Sclerotization begins in the macrochaetae of the scutellum and proceeds progressively in posterio-anterior direction over the thorax, but in the inverse direction, namely anterio-posterior, over the abdomen.

The termination of all the morphogenetic and biochemical events is indicated by the dramatic secretion of the endocuticle beginning on the tenth day, as mentioned above. Thus at the time of emergence the definite bristle is established, which is, strictly speaking, merely a specialized part of the cuticle and it should be noted that, in contrast to the remaining epidermis, the morphogenetic and biochemical processes of cuticularization have been completed before the adult insect leaves its pupa.

About the time of emergence symptoms of degeneration in bristle-forming cells become evident, such as a gradual pycnosis of the nuclei. The chromosomes now exhibit an extremely high fragility and a tendency to fuse unspecifically; the single fragments become paler and paler. These events are obviously due to decomposition of DNA and chromosomal proteins. Cytolysis of the trichogen cells is completed within the first two or three days of adult life.

B. The Footpads of Sarcophaga

The morphogenetic and biochemical capacities of the giant cells in the *Sarcophaga* footpads (Fig. 8a) closely resemble the processes of bristle formation. WHITTEN (1969) has analyzed the sequential cuticle secretion in the footpads of *Sarcophaga bullata* by electron microscopy. Some of the most important data from these investigations will be summarized below.

The general organization of the foot of *Sarcophaga* is seen in Fig. 8b. The dorsal cuticle of each pulvillus is the product of two giant cells which, after completion of growth, including rapid polytenization of their nuclei, will begin sequential deposition of individual cuticle layers on the fifth day after true pupation (rearing temperature: 25° C). The characteristic proximo-distal ridges of the cuticle surface (see Fig. 9a) are essentially built up by a corresponding accumulation of exocuticle material (layer 3; cf. Fig. 8b). The ventral sides of the pulvilli which differ from this large-faced anlage of the dorsal cuticle of each footpad by only two giant cells, are coated with thousands of tenent hairs in a beautifully precise arrangement (cf. Fig. 9b, c). Each microhair is the product of a single tenent-hair-forming cell; however, for the purpose of the present paper only the dorsal giant cells are of interest.

The growth of the dorsal giant cells is completed by the beginning of the fifth day after the "cryptocephalic" stage; now the footpads have attained their adult shape. Thereafter cuticle secretion will start with the formation of the ecdysis membrane. The next three layers (numbers 2—4) are rapidly secreted, all on day 6. On days 7—8 no further changes in the cuticle are visible. At the beginning of the ninth day a narrow band of so-called "intermediate layer" is formed. Fig. 8b diagrammatically shows the anatomical situation in a footpad on day 9, at the time when this intermediate layer is just being built up. Day 11 is characterized by the beginning of darkening in the dorsal cuticular ridges, proceeding proximo-distally and indicating

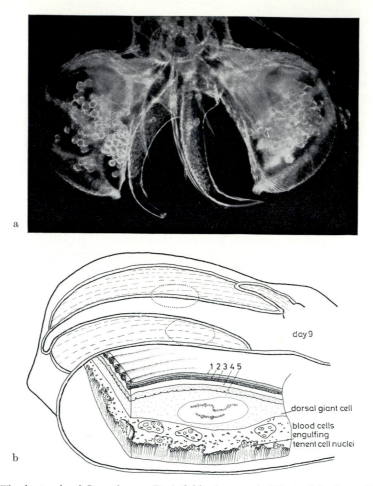

Fig. 8. The footpads of *Sarcophaga*. a Dark field micrograph (30 ×) of the foot of *Sarcophaga barbata*. The cuticular ridges on the dorsal side of each pulvillus just can be seen. b Diagrammatic representation of the anatomical situation of a footpad on the ninth day. Besides ecdysis membrane (1), four layers (2—5) of cuticle are deposited. Secretion of the last layer (endocuticle) will begin later on day 11. The left pulvillus shows a combined cross and longitudinal section; relations of magnitude are slightly distorted for better illustration of details (Modified after Whitten, 1969)

the onset of sclerotization and melanization. These biochemical alterations are said to be essentially restricted to the ridges of layer 3 ("dense exocuticle"). The developmental processes are terminated, as in the bristle-forming cells, by the dramatic deposition of the endocuticle (layer 6).

Near the time of emergence, the cytoplasm and nuclei of the giant cells are seen to degenerate. The cytolysis is obviously by the activity of hemocytes, which can be seen to be intimately associated with the giant cells at this time. Some of the most important metabolic changes related to cell growth and the formation of cuticle have been investigated in the laboratory of U. Clever (Clever, Bultmann and

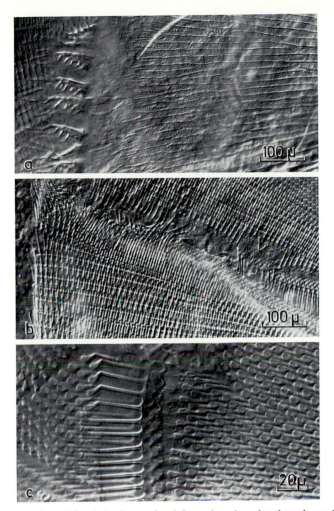

Fig. 9a—c. The adult cuticle of the footpads of *Sarcophaga* has developed specific sculpturesque structures, which, on the dorsal side a, are the product of the two giant cells, whereas on the ventral side b and c each tenent hair, with spoon-like enlargement on the tip, is formed by a single "tenent hair cell". Nomarski interference phase contrast

DARROW, 1969; BULTMANN and CLEVER, 1970). ³H-thymidine incorporation ceases between days 7 and 8 (after the white-puparium stage), thus indicating the end of chromosomal endoreduplication. In vitro application of ¹⁴C-leucine and ¹⁴C-glucose to cultured pretarsi of *Sarcophaga bullata* indicated that the major portion of the cuticle protein and chitin is synthesized between the ninth and the eleventh days .The spatial and temporal pattern of ¹⁴C-tyrosine incorporation into the dorsal footpad cuticle between days 11 and 12.5 exactly follows the schedule of tanning in situ.

In Table 1 a diagrammatic compilation of some of the most important morphogenetic and biochemical events during differentiation of the two cell systems is given. The synopsis demonstrates once more the close relationships between footpad and

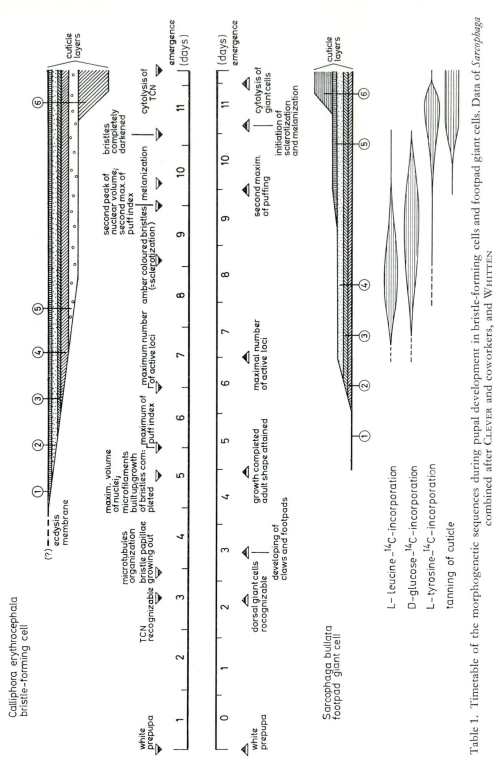

Table 1. Timetable of the morphogenetic sequences during pupal development in bristle-forming cells and footpad giant cells. Data of *Sarcophaga* combined after Clever and coworkers, and Whitten

bristle-forming cells. But there are also some differences between the two cell types: for example, the abundant production of microtubules in the growing bristle shafts, whereas in footpads microtubules are organized at a far lower rate. There are also quantitative differences in the processes of sclerotization and melanization: whereas these events in the bristles are quite significant, as indicated by the obvious changes in color, sclerotization and melanization in footpads they may be essentially restricted to the endocuticle material of the dorsal ridges.

III. The Puffing Pattern in Bristle-forming Cells and Footpad Giant Cells

A. Puffing in Trichogen Cell Nuclei

The pattern of chromosomal activity in TCN has been analyzed on the basis of a selected number of major puffs, all having more than twice the diameter of a neighboring unpuffed region at any time during development (RIBBERT, 1967). This list has been further completed in the meantime, and the diagram in Fig. 10a and b represents the alterations in the activity of 101 puffs, spread all over the chromosome complement of TCN of the scutellum. Puffs with a maximum swelling exceeding 1.5 times the diameter of unpuffed reference regions have been included in the list.

Some puffs in the early and very late periods of development could be localized only with great uncertainty; they have not been tabulated. Apart from that, a few puffs with activity periods of less than 12 h may have been overlooked. But the majority of active loci belonging to the category mentioned above have been accounted for. The smallest nucleolus organizing chromosome VI exhibits up to five bands at most, detectable in the rare cases of good synapsis, and beyond the nucleolus no further puffing structure can be identified in this chromosome. Puff size, as a criterion for RNA synthesis capacity of the corresponding activated gene loci, has been determined as the ratio between the diameter of the puffed region and that of a neighbouring segment with unchanged chromosomal architecture. For the diagrammatic representation of the puffing sequence (Fig. 10) rounded up values (each 0.5) of the activity indices have been used. The most important parameters of the puffing activities during pupal development are summarized in Figs. 11 and 12.

Prior to day 4 (after the white-puparium stage) it is so difficult to localize the puffs and measure the puffing indices that it is impossible to obtain exact data. Thus the gradual beginning of puffing activity before this point of time may be "cut off" to some extent in the diagram of Fig. 10. But the evaluation of measurements from day 4 onwards demonstrates that the important general developmental lines of the puffing can be followed from day 4 until emergence.

On day 5 the bristle-forming cells have obviously reached a peak of activity. This is reflected by the maximum total puffing indices per day, by the greatest relative number of active loci and by the highest percentage of puffs attaining their maximum expansion at this time (Figs. 11 and 12). Meanwhile, in the cytoplasm huge amounts of concentrically layered endoplasmatic reticulum, densely packed with ribosomes, become conspicuous (Fig. 13). From this time on puffing first decreases at a relatively low rate until day 8 and thereafter decays more rapidly. This can be seen in Fig. 12, where the graph of total puffing activity and that of the total number of active loci exhibits a slight shoulder by the eight day.

Fig. 10a and b. Diagrammatic representation of the puffing changes in the chromosomes of the bristle-forming cells of *Calliphora erythrocephala* (scutellum). The width of the bars indicates the average size of the puffs

If, in order to elucidate the alterations of the major puffs, the total sum of the activity indices ≥ 1.5 for each day is plotted against developmental time, one obtains a biphasic curve with a maximum at day 5 and a second, lower one at day 9, which coincides with the double-peaked curve of the changing nucleus volumes (cf. Fig. 12 and Fig. 11).

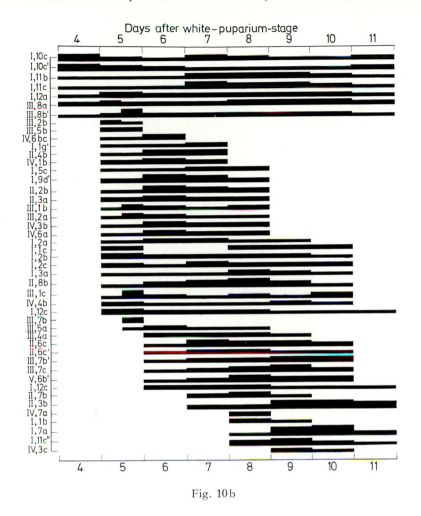

Fig. 10b

The puffs generally seem to be distributed at random along the chromosome axis, but at some points an intimate clustering of several puffs occurs within a relatively limited chromosomal section. This is true, for example, in region I, 11c, where at least four closely packed puffs develop in rapid succession, which could suggest that a puff moves along the chromosome axis in the course of time. In these relatively rare cases the size of a single puff may be slightly overestimated because in these regions swelling zones are formed by more than one puff.

All puffs represented in the diagram of Fig. 10a, b infold and disappear in a strictly sequential fashion. This sequence includes the schedule of activity as well as the time-specific degree of swelling (puffing index). Only the Balbiani ring I, 1d, active during the whole period of investigation, exhibits a certain peculiarity, in that there is considerable oscillation in the size indices for all days except day 11. Incidentally, this is the puff which develops the greatest swelling index (nearly 5). This point is attained at the time of emergence (cf. Fig. 10a, second row from the bottom).

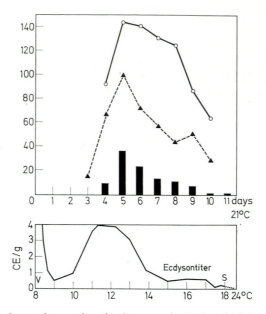

Fig. 11. Changes in the total sum of puff indices per day (○) and of the nuclear volume (▲) as a percentage of maximum volume during pupal development of *Calliphora*. The solid bars represent the number of puffs with maximal unfolding. At the bottom the changing ecdysone titer (Shaaya and Karlson, 1964) during pupal life is represented

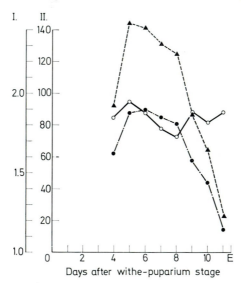

Fig. 12. Changes in the total number of active loci (●) and the total sum of puff indices per day (▲) (ordinate II). The variations of the total sum of puff indices ≧ 1.5 per number of puffs (○) show two peaks on days 5 and 9 (ordinate I); E = emergence

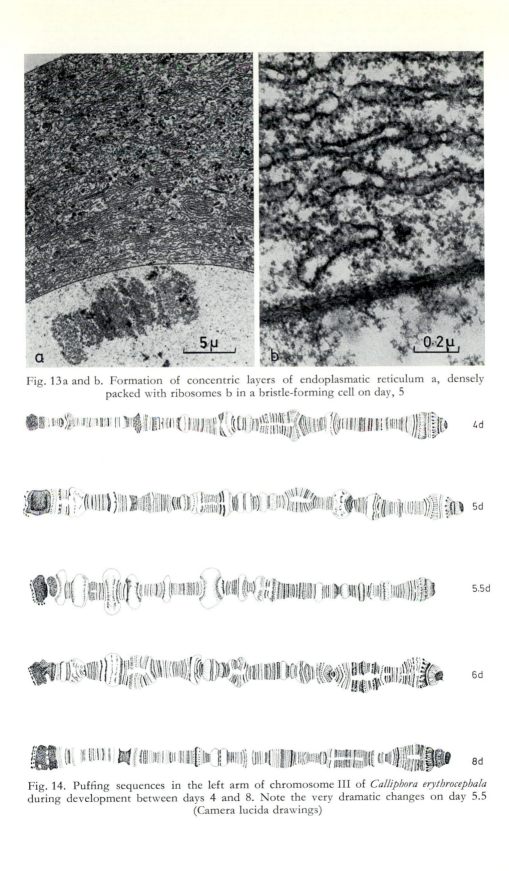

Fig. 13a and b. Formation of concentric layers of endoplasmatic reticulum a, densely packed with ribosomes b in a bristle-forming cell on day, 5

4d

5d

5.5d

6d

8d

Fig. 14. Puffing sequences in the left arm of chromosome III of *Calliphora erythrocephala* during development between days 4 and 8. Note the very dramatic changes on day 5.5 (Camera lucida drawings)

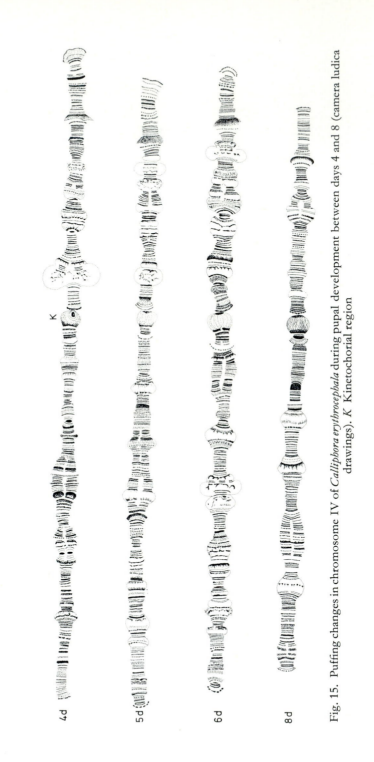

Fig. 15. Puffing changes in chromosome IV of *Calliphora erythrocephala* during pupal development between days 4 and 8 (camera ludica drawings). *K* Kinetochorial region

Figs. 14 and 15 show two examples of the dramatic changes in puffing pattern between days 4 and 8. The puffing timetable of the bristle-forming cells from anterior regions of the thorax seems to be displaced compared with that of the scutellar bristle-forming cells, in that the latter precedes that of the pro- and mesothorax, a fact which is correlated with the faster development of the scutellar bristles. Similarly,

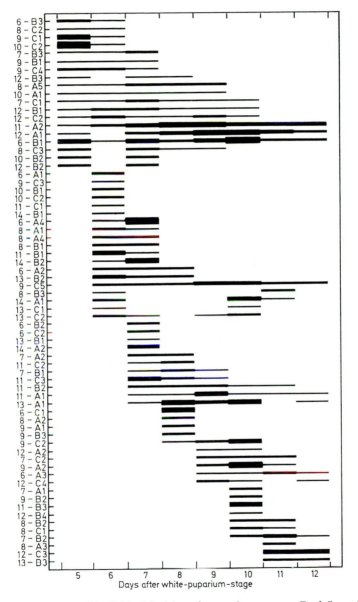

Fig. 16. Puffing changes of individual loci in polytene chromosome B of *Sarcophaga bullata* footpads during pupal development. The puffs were studied at 24 h intervals. The width of the bars indicates the average size of the puffs (BULTMANN and CLEVER, 1969)

Trepte (1971) demonstrated in *Sarcophaga* that a delay (more than 1 day at 21° C breeding temperature) of the whole puffing schedule in posterior-anterior direction coincides with a corresponding temporal gradient of bristle development. A delay of sequential puffing also occurs in the nuclei of the intimate associates of bristle-forming cells that are the tormogen cells, in addition to the formation of an increasing number of cell-specific puffs in both types of cells during the second half of pupal development.

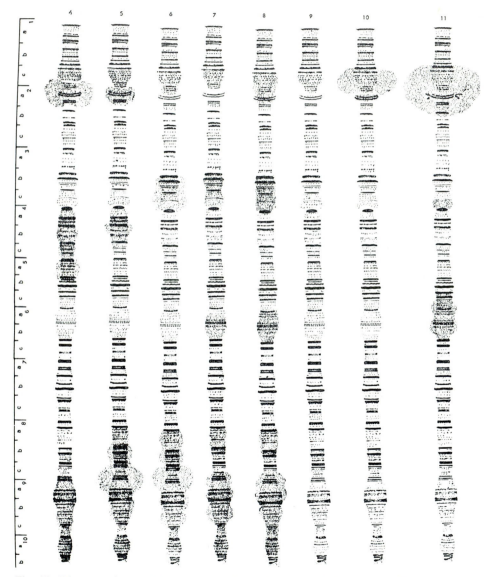

Fig. 17. Diagram of the puffing sequences in the long arm of chromosome E of *Sarcophaga bullata* during pupal development between days 4 and 11 after the cryptocephalic stage (Whitten, 1969)

B. Puffing in Dorsal Giant Cells of the Footpads

The puffing pattern of the giant footpad cells in *Sarcophaga bullata* has been presented in two papers (WHITTEN, 1969b; BULTMANN and CLEVER, 1969). WHITTEN'S investigation covered a select group of major puffs distributed all over the complement of the polytene chromosomes. BULTMANN and CLEVER have carefully analyzed the puffing schedule of all cytologically detectable puffs occurring in chromosome B.

As in *Calliphora*, the puffs can be localized and their size determined from the time shortly before the completion of growth (5 days after the white-puparium stage) until degeneration of the nuclei at the time of emergence (12 days). The diagram of

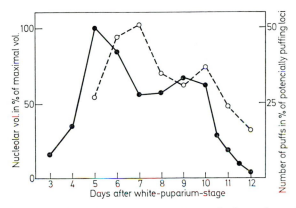

Fig. 18. Changes in the number of puffs ($\circ$) and the nucleolar volume ($\bullet$) (planimetric determination) during pupal development (BULTMANN and CLEVER, 1969)

Fig. 16 represents the alterations in size of 69 individual puffs examined by BULTMANN and CLEVER (1969), and Fig. 17 gives an example of WHITTEN'S graphic description of puffing changes. The outstanding feature of the functional pattern is the striking sequential order of appearance and disappearance of all puffs investigated. The quantitative pattern, as judged by puff sizes, is likewise determined exactly in the course of development. In general, there is one maximum of active loci on days 6 and 7 and a second, less prominent one on day 10 (cf. Fig. 18). A list of some 30 major puffs (Table 2) denoting the sequence of their most marked puffing, also shows days 6 and 7 (days 5 and 6 in WHITTEN'S staging) as those with the greatest number of fully developed puffs. This biphasic curve of the puffing activity is paralleled by a similar graph of the changing nucleolar volume; but the latter is shifted by one day towards earlier developmental stages (Fig. 18). It seems as if a rise in nucleolar activity precedes that of total puffing (BULTMANN and CLEVER, 1969).

Pupae of the same chronological age exhibit a very similar puffing pattern. A shift of the entire schedule naturally occurs when there is a delay or an acceleration of the general developmental rate. This can be verified by investigating samples of starved pupae, which, owing to their lower fresh weight, develop faster than well-fed ones, or by analyzing pupae reared at higher temperatures.

Nevertheless, the "outer" and "inner" pulvilli cells of an individual may differ characteristically in the timing of the activity of some individual loci relative to the

Table 2. Table of the major puffs in the chromosomes of the footpads of *Sarcophaga bullata*. X denotes the period of maximal unfolding (Whitten, 1969)

Puff location	D 4	D 5	D 6	D 7	D 8	D 9	D 10	D 11	Figures
A-5-c						X	X	X	2b, c, d; 12b
A-6-a						X			2d, 12b
A-6-b	X	X							2a
A-8-b/c				X					14a (Trich)
A-9-b		X		X					7 (DNA puff)
B-5-b			X						
B-7-a/b						X	X		10b, c
B-7-c						X			10c
B-8-a						X			10c
B-8-b		X	X						10a
B-8-c	X	X							4a (Balbiani ring)
B-9c/10a			X						9c, 10a
B-10-b			X						4b
B-12-a, -12-b			X						4b
C-2-a						X			
C-2-b	X								13b (Balbiani ring)
C-9-b/c/10a		X							3c
C-10-b/c						X	X		2c, 3b, 12b
C-11-b	X	X							3c (Balbiani ring)
D-1-b/c		X	X						11a
D-2-a		X	X						11a
D-3-c			X						11a
D-5-a/b/c			X						11a
D-6-b/c		X	X						11a
D-7c/9b		X							3c
D-10-a					X				8
E-1-c								X	5c, 6b, 12a
E-2-a	X								5a (Balbiani ring)
E-3-b/c			X						6a
E-8c/9a		X							3c
E-12-b			X						3c
E-13-b/c			X						3c
E-16-b/c		X							3c

rest of the puffing sequences, as has been demonstrated by Bultmann and Clever (1969). At some gene positions, puff unfolding in the outer giant cells of the footpads precedes that in the inner cells but persists longer in the inner cells than in the outer ones.

A general shift of the entire puffing sequences is indicated by the general delay of puffing changes in the metathoracic footpads as compared with those of the pro- and mesothorax.

IV. Discussion

Both the bristle-forming apparatus and the growing footpads are cell systems which are closely connected with the developmental events during pupal metamorphosis. They are, furthermore, „Kleinorgane", a term which is applied to small integumental structures such as bristles, hairs, scales, chemosensilla etc. (HENKE, 1953), with a comparatively simple morphogenesis.

At relatively early stages of pupal differentiation, even before the proper growth period of the forming cells is completed, the size of the polytene chromosomes and the quality of their banding pattern are sufficient for puffing analysis, that is, at the latest from the 4th or 5th day on. This is of great importance because all the essential developmental events during the formation of the bristles or footpads until their final state at the time of emergence take place in cells with easily analyzable polytene chromosomes.

The close relationship of the bristle- and footpad-forming cells to the pupal metamorphic events is demonstrated by a series of cellular functions which have been shown to change sequentially with bristle formation and leg development. Thus the outstanding feature of these epidermal cells is the *fixed sequence of defined stages of differentiation* during pupal development. In contrast, the salivary glands of *Chironomus*, for example, are engaged in a relatively constant secretory function throughout larval life and hardly undergo any developmental alterations. Accordingly, the majority of puffs do not change their activity in relation to development (CLEVER, 1962; 1963).

In the case of differentiating pupal epidermal cells, closer attention can now be devoted to the question of the nucleocytoplasmic relationships within a type of cells which pass through different specific states of synthesis activity as they approach adulthood by metamorphosis. The fact that puffing sequences in bristle-forming and footpad giant cells runs so strictly parallel to the sequence of morphogenetic and biochemical events during pupal development makes it obvious that the sequential initiation of new cell functions in these epidermal cells is directly dependent upon a correspondingly changing genomic activity, as expressed in the strictly sequential changes of the puffing pattern of the polytene chromosomes.

The developmental state on the 5th day in trichogen cells, or on the 7th day in footpads, seems to represent a turning point, as characterized by the dramatic unfolding of puffs, some only for a limited period of perhaps less than twelve hours (cf. Figs. 10a, b and 16). Many cytological parameters attain their maximum values at this time (cf. Figs. 11, 12 and 18), and these events coincide with the main secretory activities of the bristle-forming and footpad giant cells during cuticle deposition.

The immediacy of genomic control seems also to be indicated by the delay of the whole puffing schedule, which is observed whenever there is a corresponding shift in the onset of sclerotization and melanization in different regions of the thorax and abdomen. The concept of immediate genomic control is supported particularly by the finding that certain biochemical steps of synthesis involved in the tanning process, e.g. the incorporation of tyrosine, depend upon concurrently synthesized RNA and protein. On the other hand, incorporation of dopa and of dopamine into the cuticle has been shown to be insensitive to the application of cycloheximide and actinomycin-D respectively, which would suggest that perhaps only a few steps of a particular biochemical pathway are immediately controlled by the genome (CLEVER, BULTMANN and DARROW, 1969). Furthermore, some of the processes of cuticle formation and tanning require RNA synthesis for their initiation, as demonstrated by their actinomycin sensitivity, but seem to become independent once they have started, for they are then insensitive to actinomycin, but still sensitive to cycloheximide (Clever, personal communication). Thus, once the newly synthesized m-RNA, presumable initiating a particular developmental event, has attained a certain level in the cytoplasma, it can probably maintain protein synthesis and would seem to be rather stable.

The hormone ecdysone has been shown to occupy the key position in regulating the metamorphic events of larval-pupal molt. On the cytogenetic level it has been demonstrated that the puffing sequence in salivary gland chromosomes of *Chironomus* and *Drosophila* is induced by an increase in the concentration of ecdysone in the hemolymph at the end of the last larval stage (Clever, 1961; Becker, 1962; Berendes, 1967; Ashburner, this volume). The stimulating effect of the molting hormone and the biochemical events involved in cuticle sclerotization have been analyzed in the formation of the puparium of *Calliphora*.

Cuticle hardening is achieved by quinone tanning, which process probably meshes cuticle protein chains with each other or with chitin. N-acetyl-dopamine has been found to be the real sclerotizing substance, which — after an ecdysone-induced switchover of the tyrosine metabolism — accumulates in the hemolymph of the larvae shortly before puparium formation (Karlson and Sekeris, 1962a, b; Karlson, 1965; Karlson and Peters, 1965).

Unlike the remaining epidermis, which is tanned and melanized spontaneously shortly after emergence, footpads, and to a far greater extent bristles, sclerotize and darken during pupal life, and all morphogenetic and biochemical events for the organization of the definitive bristle and footpad cuticle have been completed by emergence.

Shaaya and Karlson (1965) have measured the changing ecdysone titer during the postembryonal development of Calliphora. In addition to a peak at the time of puparium formation, a second, lower maximum is observed between the 4th and 5th days. This peak slightly precedes the period of most intensive chromosomal activity in the bristle-forming cells of *Calliphora* and the same would be true if the measurements of the hormone titer in *Calliphora* were extrapolated to *Sarcophaga*. The decline of general puffing activity in trichogen cell nuclei in principle follows the decline of the ecdysone titer. It thus seemed likely that the majority of the puffs were directly controlled by ecdysone.

We carried out a series of experiments in our laboratories, injecting ecdysone[2] of various concentrations and different incubation times both before and after the natural peak at days 4—5; but, so far, we have not been able to detect experimental puff induction.

These results are quite surprising. We had expected that ecdysone would exert a particularly strong effect on these epidermal cells, where the biochemical events occurring during sclerotization were thought to be the same as those observed in epidermal cell during larval-pupal molt under the control of ecdysone. It looks as if these biochemical processes in the developing bristle cells are autonomously controlled. The same state of affairs is indicated by the autonomy of individual footpad cells in the timing of their puff activity (Bultmann and Clever, 1969).

The subdivision of a continuous developmental process into stages is always arbitrary. Nevertheless, some periods in the life cycle of the bristle-forming cell can

---

2 Kindly furnished by Prof. P. Karlson.

be distinguished by the onset and termination of remarkable morphological and cytological changes.

The first period is characterized by intensive cell growth, by the synthesis and organization of a cytoskeleton of microtubules, which probably play an important role in the morphogenesis of the bristle. With the formation of microfilaments and the anlage of concentric layers of rough endoplasmatic reticulum round the cell nucleus, this growth period is completed on about the 5th day. The period of greatest secretory activity in the cell coincides with the most intensive overall puffing. The deposition of the individual cuticle layers — except the endocuticle — seems to be finished by about the 8th day at the latest. At this time, the beginning of sclerotization is indicated by a gradually increasing amber coloration of the bristles, followed by melanin pigmentation. The rapid deposition of the endocuticle on the 11th day terminates the "work" of the bristle-forming cells which start to degenerate at the time of emergence. Even cytolysis seems to be a self-regulated process: Most of the puffs have ceased their activity shortly before emergence, so that the small number of late puffs are all the more conspicuous. For example, the largest puff ever developing in the trichogen cell genome, Balbiani ring I, 1d (cf. Fig. 10a, second position from the bottom) attains its maximum dimension at the time of emergence.

Any attempt to associate certain individual puffs with specific cell functions in bristle-forming cells or footpad giant cells would be highly speculative in view of the present state of our knowledge. But we are justified in predicting that, among the sequentialy changing puffs, just the major ones will be found to be engaged in the genetic control of the striking phenotypic characters of bristles or footpads at distinct periods of pupal development.

To establish this point, one must look for suitable mutations allowing direct genetic correlation between certain puffs and distinct morphological or biochemical features.

References

ASHBURNER, M.: Patterns of puffing activity in the salivary gland chromosomes of Drosophila. I. Autosomal puffing patterns in a laboratory stock of Drosophila melanogaster. Chromosoma (Berl.) 21, 398—428 (1967).

BAUDISCH, W.: Spezifisches Vorkommen von Carotinoiden und Oxyprolin in den Speicheldrüsen von Acricotopus lucidus. Naturwissenschaften 47, 498—499 (1960).

— Synthese von Oxyprolin in den Speicheldrüsen von Acricotopus lucidus. Naturwissenschaften 48, 56 (1961).

— PANITZ, R.: Kontrolle eines biochemischen Merkmals in den Speicheldrüsen von Acricotopus lucidus durch einen Balbiani-Ring. Exp. Cell Res. 49, 470—476 (1968).

BECKER, H. J.: Die Puffs der Speicheldrüsenchromosomen von Drosophila melanogaster. I. Beobachtungen zum Verhalten des Puffmusters im Normalstamm und in zwei Mutanten, giant and lethal giant-larval. Chromosoma (Berl.) 10, 654—678 (1959).

— Die Puffs der Speicheldrüsenchromosomen von Drosophila melanogaster. II. Mitt. Die Auslösung der Puffbildung, ihre Spezifität und ihre Beziehung zur Funktion der Ringdrüse. Chromosoma (Berl.) 13, 341—384 (1962).

BEERMANN, W.: Chromosomenkonstanz und spezifische Modifikationen der Chromosomenstruktur in der Entwicklung und Organdifferenzierung von Chironomus tentans. Chromosoma (Berl.) 5, 139—198 (1952).

— Nuclear Differentiation and Functional Morphology of Chromosomes. Cold Spring Harb. Symp. quant. Biol. 21, 217—231 (1956).

— Chromosomal differentation in insects. In: RUDNICK, D. (Ed.): Developmental Cytology, pp. 83—104. New York: Ronald Press Comp. 1959.

178 D. Ribbert:

Beermann, W.: Ein Balbianiring als Locus einer Speicheldrüsenmutation. Chromosoma (Berl.) **12**, 1—25 (1961).
— Operative Gliederung der Chromosomen. Naturwissenschaften **52**, 365—375 (1965).
Berendes, H.D.: Salivary gland function and chromosomal puffing patterns in Drosophila hydei. Chromosoma (Berl.) **17**, 35—77 (1965).
— The hormone ecdysone as effector of specific changes in the pattern of gene activities of Drosophila hydei. Chromosoma (Berl.) **22**, 274—293 (1967).
Bultmann, H., Clever, U.: Chromosomal control of footpad development in Sarcophaga bullata. I. The puffing pattern. Chromosoma (Berl.) **28**, 120—135 (1969).
— — Chromosomal control of footpad development in Sarcophaga bullata. II. Cuticle formation and tanning. Develop. Biol. **22**, 601—621 (1970).
Clever, U., Bultmann, H., Darrow, J.M.: The immediacy of genomic control in polytenic cells. In: Hanly, E.W. (Ed.): Problems in Biology, pp. 403—423. Salt Lake City: University of Utah Press 1969.
— Genaktivitäten in den Riesenchromosomen von Chironomus tentans und ihre Beziehung zur Entwicklung. I. Genaktivierung durch Ecdyson. Chromosoma (Berl.) **12**, 607—675 (1961).
— Karlson, P.: Induktion von Puffveränderungen in den Speicheldrüsenchromosomen von Chironomus tentans durch Ecdyson. Exp. Cell. Res. **20**, 623—626 (1960).
Goldberg, E., Whitten, J., Gilbert, L.: Changes in soluble proteins during foot-pad development in Sarcophaga bullata. J. Insect Physiol. **15**, 409—420 (1969).
Grossbach, U.: Chromosomen-Aktivität und biochemische Zelldifferenzierung in den Speicheldrüsen von Camptochironomus. Chromosoma (Berl.) **28**, 136—187 (1969).
Henke, K.: Über Zelldifferenzierung im Integument der Insekten und ihre Bedingungen. J. Embryol. exp. Morph. **1**, 217—226 (1953).
Karlson, P.: Biochemical studies of ecdysone control of chromosomal activity. J. cell. comp. Physiol. **66**, 69—76 (1965).
— Sekeris, C.: N-Acetyldopamine as sclerotizing agent in the insect cuticle. Nature **195**, 183—184 (1962).
— — Zum Tyrosinstoffwechsel der Insekten; IX. Kontrolle des Tyrosinstoffwechsels durch Ecdyson. Biochem. biophys. Acta (Amst.) **63**, 489—495 (1962).
— Peters, G.: Zum Wirkungsmechanismus der Hormone. IV. Der Einfluß des Ecdysons auf den Nucleinsäurestoffwechsel von Calliphora-Larven. Gen. Comp. Endocr. **5**, 252—259 (1965).
Lawrence, P.A.: Development and determination of hairs and bristles in the milkweed bug, Oncopeltus fasciatus. J. Cell Sci. **1**, 475—498 (1966).
Lees, A.D., Picken, L.E.R.: Shape in relation to fine structure in the bristles of Drosophila melanogaster. Proc. Roy. Soc. **B132**, 396—432 (1945).
Locke, M.: The structure of an epidermal cell during the development of the protein epicuticle and the uptake of molting fluid in an insect. J. Morph. **127**, 7—39 (1969).
Mechelke, F.: Reversible Strukturmodifikationen der Speicheldrüsenchromosomen von Acricotopus lucidus. Chromosoma (Berl.) **5**, 511—543 (1953).
— The timetable of physiological activity of several loci in the salivary gland chromosomes of *Acricotopus lucidus*. Proc. X. Int. Congr. Genetics **1958 II**, 183.
Overton, J.: The fine structure of developing bristles in wild type and mutant Drosophila melanogaster. J. Morph. **122**, 367—380 (1967).
Panitz, R.: Innersekretorische Wirkung auf Strukturmodifikation der Speicheldrüsenchromosomen von Acricotopus lucidus. (Chironomidae). Naturwissenschaften **47**, 389 (1960).
— Hormonkontrollierte Genaktivitäten in den Riesenchromosomen von Acricotopus lucidus. Biol. Zbl. **83**, 197—230 (1964).
— Funktionelle Veränderungen an Riesenchromosomen nach Behandlung mit Gibberellinen. Biol. Zbl. **86** (Suppl.) 147—156 (1967).
Pelling, C.: Chromosomal synthesis of ribonucleic acid as shown by the incorporation of uridine labeled with tritium. Nature (Lond.) **184**, 655—656 (1959).
— Ribonukleinsäuresynthese der Riesenchromosomen. Autoradiographische Untersuchungen an Chironomus tentans. Chromosoma (Berl.) **15**, 71—122 (1964).

RIBBERT,D.: Die Polytänchromosomen der Borstenbildungszellen von Calliphora erythrocephala unter besonderer Berücksichtigung der geschlechtsgebundenen Heterozygotie und des Puffmusters während der Metamorphose. Chromosoma (Berl.) **21**, 296—344 (1967).

SHAAYA,E., KARLSON,P.: Der Ecdysontiter während der Insektenentwicklung. II. Die postembryonale Entwicklung der Schmeißfliege Calliphora erythrocephala. J. Insect Physiol. **11**, 65—70 (1964).

SLIZYNSKI,B.M.: Functional changes in polytene chromosomes of Drosophila melanogaster. Cytologia (Tokyo) **29**, 330—336 (1964).

TREPTE,H.: Vergleichende Untersuchungen der Chromosomenaktivität während der Differenzierung der polytänen Zellen des Borstenbildungsapparates von Sarcophaga. In preparation.

WHITTEN,J.M.: Giant polytene chromosomes within hypodermal cells of developing footpads. Science **143**, 1437—1438 (1964).

— Coordinated development in the fly foot: Sequential cuticle secretion. J. Morph. **127**, 73—104 (1969).

— Coordinated development in the foot pad of the fly Sarcophaga bullata during metamorphosis: changing puffing patterns of the giant cell chromosomes. Chromosoma (Berl.) **26**, 215—244 (1969).

The Control of Puffing in Drosophila hydei

HANS D. BERENDES

Department of Genetics, University of Nijmegen, Nijmegen

I. Introduction

In *Drosophila hydei*, as in other Dipteran species, numerous puffs can be observed along the polytene chromosomes in larval and in some adult cell types. Each of these puffs includes a section of the DNA of the genome from which RNA is transcribed. Because, in the living cell, transcription of chromosomal DNA may be regarded as an expression of gene activity, the pattern of puffs in any particular type of cell should represent the pattern of active genes. The correctness of this conclusion, however, largely depends upon the criteria by which a puff is defined.

1. The Pattern of Active Genes

Autoradiographs prepared from *Drosophila* polytene cells which were exposed during a short period to radioactive RNA precursors reveal incorporation of the precursor into ribonuclease-digestible material at numerous chromosome regions. The level of incorporation varies greatly from one region to another (Fig. 1). The nucleolus and regions which display the typical morphology of a puff, an increased chromosome diameter in combination with a dispersed banding pattern, show in general relatively high levels of incorporation. On the other hand, various chromosome regions can be found which, though incorporating RNA precursors, fail to show the characteristic morphology of a puff. Whereas this failure could merely be a consequence of a much lower level of transcription as compared with the major puffs, it raises the problem as to why these regions escape from the general scheme according to which the pattern of puffs should correspond with the pattern of active genes.

It appeared however, that these nonpuffed, active regions have features in common with morphologically distinct puffs. Application of the acidic fast-green staining procedure (ALFERT and GESCHWIND, 1953) revealed discrete transverse green bands in regions which cannot be defined as puffs using morphological criteria. These bands, which are indicative for dye binding to non-histone proteins, are located in regions active in RNA synthesis. Some of these regions, furthermore, reveal at the submicroscopic level the presence of particulate material distributed over a minute transverse section of the chromosome. The individual particles resemble those regularly observed in active puffs, with respect to their electron density as well as their diameter ($\sim$ 300 Å) (Fig. 2).

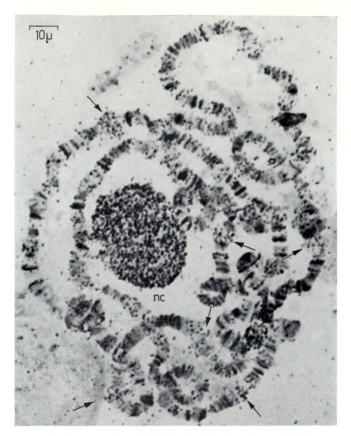

Fig. 1. The pattern of incorporation of tritiated uridine in a late third instar salivary gland nucleus of *Drosophila hydei* 20 min after the injection of 1 μC per larva. In addition to the obviously puffed regions (arrows), the uridine is incorporated at many sites which fail to show an evident puff morphology. nc: nucleolus

Regions displaying the features just described would be identified with the light microscope as interband regions which show no obvious sign of puffing in a morphological sense. It seems appropriate to define these regions, which display, apart from the morphology, all other characteristic features of puffed regions, as "micropuffs".

2. Puffing Patterns

Thus far, most investigations on puffing patterns of *Drosophila* have been more or less exclusively based upon morphological criteria (BECKER, 1959; BERENDES, 1965a; ASHBURNER, 1967a; 1969a). These patterns, therefore include only those genome sites which have attained a certain minimum level of activity (size).

Nevertheless, it was revealed by using morphology as the only criterion that the puffing pattern of a particular cell type varies with development (BECKER, 1962; BERENDES, 1965a) and responds to changes in the extracellular environment (KROEGER, 1960). In particular, the possibility of effecting a controlled change in the

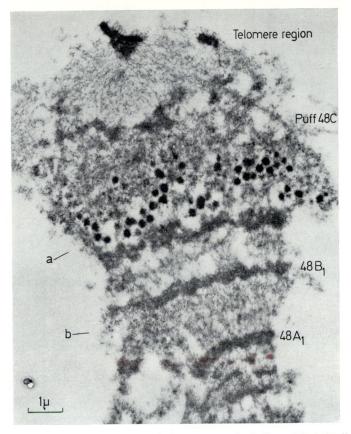

Fig. 2. Electron micrograph of the tip of chromosome 2 of *Drosophila hydei* displaying an experimentally induced puff in region 48 C and a "micropuff" in region 48 A (arrow b). In region 48 C (top) and 48 A, ribonucleoprotein particles are present. Aggregation of these particles has occurred in puff 48 C (arrow a); this is the *only* puff in *Drosophila hydei* in which this phenomenon has been observed

extracellular environment by experimental means, with a consequent definite change in the pattern of active genes, seemed promising for a more detailed study on the puffing phenomenon and its control.

3. Experimental Puff Induction

At present, a great variety of agents is known to cause definite agent-specific responses at the genome level. Among these agents, some are regularly involved in physiological events occurring during normal development. The best-known examples are the steroid molting hormone ecdysone (CLEVER and KARLSON, 1960; BURDETTE and ANDERSON, 1965; BERENDES, 1967; CROUSE, 1968; BERENDES and THIJSSEN, 1971; ASHBURNER, 1971) and the juvenile hormone (LEZZI and GILBERT, 1969; LAUFER and HOLT, 1970). With regard to other agents, such as a sudden rise in temperature (RITOSSA, 1962, 1964; BERENDES, VAN BREUGEL, and HOLT, 1965;

van Breugel, 1966; Ashburner, 1970a), gibberellic acid (Panitz, 1968; Baudisch and Panitz, 1968; Alonso, 1971), certain metal ions (Kroeger, 1964, 1968), ribonuclease (Ritossa et al., 1965), oxytetracycline (Serfling, Panitz, and Wobus, 1969) actinomycin D (Kiknadze, 1965), dicyandiamide (Mukherjee, 1968), thiocyanate (Sakharova, Rapaport, Beknazar'yants and Nikiforov, 1971) and tryptophan (Fedorov and Milkman, 1964), it is uncertain whether their effect upon the pattern of genome activity observed under experimental conditions has any relevance for cellular metabolism during normal development.

Irrespective of whether or not these agents affect cellular physiology during normal development, it is evident that they do induce changes in the puffing pattern. These agents are thus appropriate tools for the investigation of some pertinent questions concerning the regulatory devices participating in the control of gene activity (puffing) in multicellular organisms.

In the first instance, these agents may be used for investigations on a number of aspects of the puffing phenomenon itself in order to define, at the chromosomal level, the events involved in puff formation and puff regression. This, consequently, will raise questions with regard to the qualitative and quantitative control of puffing. Furthermore, because the mechanisms for qualitative as well as quantitative control of puffing should be related to the stimulus on the one hand, and to the basic metabolic situation of the responding cell on the other hand, experimental induction under controlled conditions may be applied to investigate these relationships in more detail.

In an attempt to approach some of these questions, attention will be focused on the results obtained with two devices for experimental induction of changes in the puffing pattern of *Drosophila hydei*: a temperature treatment, and the steroid hormone ecdysone.

II. Experimental Induction of Puffs in Drosophila hydei

In all experimental approaches to be described, salivary gland chromosomes of mid third instar larvae have been used for identification of changes in the puffing pattern. The basic puffing pattern of this cell type and its changes in the course of normal development have been investigated in detail throughout the third larval instar and the prepupal period (Berendes, 1965a and unpublished).

A. Puffs Related to the Respiratory Metabolism

1. Puff Induction by Temperature Treatment

Temperature treatments were performed in vivo by transferring larvae from 25° to 37° C, or in vitro with isolated glands incubated in Ringer's solution by raising the temperature over the same range. These treatments induce new puffs at six loci of the genome (Berendes, van Breugel and Holt, 1965; van Breugel, 1966) (Fig. 3). It should be mentioned that, though the effect of a temperature treatment is identical under in vivo and in vitro conditions, in the experiments with isolated glands some puffs appear which are found neither in the in vivo treated, nor in control (untreated) larvae of the same age. These puffs appear to occur regularly in explanted glands upon incubation.

The time between the onset of a temperature treatment and the first visible (morphological or cytochemical) indication of puff development is extremely short

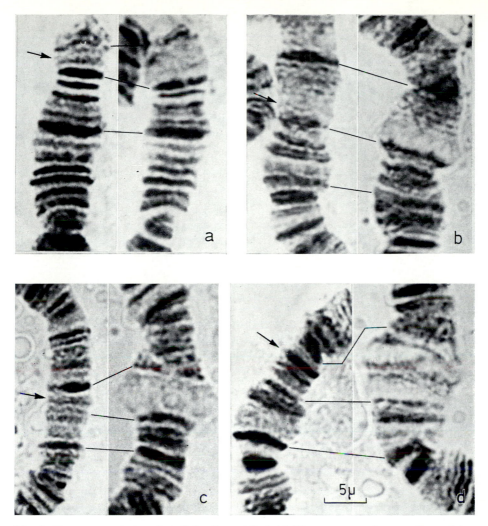

Fig. 3. Temperature-induced puffs in *Drosophila hydei*. The most prominent puffs resulting from a 15-min temperature treatment in vivo are 48C (a); 36A (b); 81B (c) and 32A (d)

(3—5 min) (BERENDES, 1968; VAN BREUGEL, 1966). The newly arising puffs attain their maximum size within 30 min. The changes in the puffing pattern induced by a temperature treatment ("temperature puffs") are reversible. Upon returning the larvae from 37° to 25°, the induced puffs regress completely within 30—40 min. Regression also occurs when the larvae are kept for longer than 60 min at 37°. The puffs then begin to regress at some time between 60 to 80 min after the onset of the treatment.

Although the six loci, four of which are shown in Fig. 3, consequently start puffing if the animals are submitted to a temperature treatment, it appears that puff formation at these loci is not restricted to this treatment alone. Puffing of the same loci may be observed under conditions of constant temperature (25°) in normal development coincidentally with the changes in the puffing pattern associated with puparium formation. Under these circumstances the "temperature" puffs are generally smaller than following experimental induction. It may also be mentioned that, in contrast to the situation after experimental induction, naturally occurring "temperature" puffs vary considerably in size. Temperature puffs may also be observed in animals grown under crowded conditions (at 25° C). The mere fact that these puffs do occur in normally developing larvae indicates that the activity of these loci may be, under certain circumstances, relevant for the metabolism of the cell.

So far, however, relatively little is known about the possible function of the RNA synthesized in these puffs. A similar ignorance exists with regard to the mechanism(s) responsible for the induction and control of puff formation at these loci.

2. Metabolic Pathways Related to the Temperature-specific Genome Response

A variety of experiments have recently been conducted in order to establish the pathway(s) of cellular metabolism which are affected by a temperature treatment. Previous studies have indicated that the oxygen or energy metabolism of the cells is affected by the temperature treatments.

Observations on salivary glands of *Drosophila busckii* incubated in a medium containing uncouplers of the oxidative phosphorylation, such as dinitrophenol (DNP, 10^{-3}M) or sodium salicylate (10^{-2}M), revealed that these compounds induce exactly the same changes in the puffing pattern as temperature treatments (Ritossa, 1962, 1964). A repetition of these experiments with salivary glands of *D. hydei* essentially confirmed these results. On account of the effect of these uncouplers of the oxidative phosphorylation, it could be argued that the "temperature" puffs arise as a consequence of a deficiency in cellular ATP supply. A sudden rise in environmental temperature would stimulate metabolism and raise the energy requirements suddenly to an extent which cannot instantly be met by the metabolic pathways providing ATP. In its simplest form, this interpretation seems to encounter serious difficulties. It was established that larvae or isolated glands of *D. hydei* which are kept under anaerobic conditions (CO_2 or N_2 atmosphere) for various periods, fail to develop "temperature" puffs although the treatment as such should effectively reduce the cellular ATP level. If the larvae or isolated glands become exposed to fresh air or oxygen after being submitted to an anaerobic treatment, the "temperature" puffs develop. The rate of development of the puffs as well as the length of the period over which they are active, was found to be related to the length of the period of anaerobiosis. Whereas after a one hour CO_2 treatment, the puffs develop within the first 10 min of recovery in air and remain active over a period of approximately 50 min, after a 3 hour CO_2 treatment it takes 40—50 min of recovery before the puffs have been developed and a period of 120—150 min before they have regressed again (van Breugel, 1966).

Another series of experiments did not support the suggestion that the "temperature" puffs should develop in response to a deficiency in cellular ATP. Isolated glands which were incubated in a medium containing oligomycin (saturated solution), a substance which also inhibits oxidative phosphorylation, failed to develop "temperature" puffs.

It was these results that directed the study of the mechanism triggering the induction of "temperature" puff formation to a search for a relationship with respiratory chain reactions.

3. Respiratory Chain Reactions Involved in "Temperature" Puff Induction

In order to exclude the possibility that metabolic pathways other than those involved in terminal respiration are responsible for the induction of "temperature" puff activity, LEENDERS (1972) provided isolated glands incubated in an appropriate medium, with a variety of substrates resulting from glycolysis or from lipid metabolism. If these substances, Na-pyruvate and Na-acetate, were limiting factors in the sense that an insufficient availability of these substrates should evoke the induction of the development of "temperature" puffs, their supply to the medium should result in the absence of puff formation when a puff inducing stimulus is applied. However, a temperature treatment or addition of DNP *did* induce the specific genome response in the presence of these compounds.

On the other hand, if substrates of the citric acid cycle, e.g. Na-malate and Na-succinate are supplied to the medium, the genome response following a temperature treatment or administration of DNP is effectively reduced or even completely absent.

Further arguments for a relationship between terminal respiration reactions and the activity of the "temperature" sensitive genome loci were obtained from studies on the effect of substances which specifically inhibit certain reactions in the respiratory chain. Administration of rotenone or amytal, potent inhibitors of the H-transfer between NADH and CoQ, to isolated salivary glands, did induce the appearance of the "temperature" puffs without a temperature treatment. Also the addition of anti-mycin A or 2-heptyl-4-hydroxyquinoline-N-oxide (HQNO), substances which suppress the electron transfer between cytochrome b and cytochrome c (MAHLER and CORDES, 1969), to isolated salivary glands were equally effective (LEENDERS, 1972).

It seems reasonable to expect that by the inhibition of a definite step in the reaction chain, all intermediates after this block shift to a more oxidized state, whereas the intermediates of the chain before the block become in a more reduced state. It could be supposed that the oxidized state of a part of the respiratory chain functions as a signal which, by some unknown mechanism, is responsible for the induction of puff formation. Support for this suggestion was found in the effect of substances which may act as "artificial" hydrogen acceptors. Incubation of the glands with menadione (Vitamin k_3) or oxymethylene blue, both of which could produce a more oxidized state of the respiratory chain, did induce "temperature" puffs.

On account of the results presented so far, it seems that the genome response to changes in the respiratory metabolism occur as a consequence of a change in redox state of the respiratory chain rather than as a result of an ATP deficiency. This idea, however, depends strongly on the observation that the inhibitor oligomycin fails to induce the appearance of the "temperature" puffs.

Some data do not support this hypothesis. For example, it was found that a supply of exogenous ATP to isolated salivary glands effectively reduced the effect of respiration stimulating treatments e.g. a temperature treatment or DNP. The appearance of "temperature" puffs was significantly delayed or inhibited under these consitions. With regard to the negative effect of oligomycin, it could be argued that its interference with the production of ATP under the experimental conditions used, does not reduce the cellular ATP level to such an extent that the appropriate genome response becomes initiated.

At the present, a definite answer to the question as to whether the redox state of the respiratory chain or a deficiency in cellular ATP (probably as a consequence of the changes in the respiratory metabolism) is responsible for the induction of the temperature puffs, cannot be given. On the other hand, the experiments here described have provided results which strongly suggest that changes in the mitochondrial metabolism may evoke a particular response of nuclear genes. In this context, it seems also interesting that following the activity of the "temperature" puffs, the quantity of the enzyme NADH-dehydrogenase increases, an increase which can be inhibited by actinomycin D and cycloheximide (Leenders and Beckers, 1972).

It may be suggested that a further unravelling of the complex system just described, may eventually reveal not only the factor, or factors, which are responsible for the initiation of the activity of particular genes, but also the significance of this activity for cellular metabolism.

B. Puffs Related to Ecdysone

1. Puff Induction by the Molting Hormone, Ecdysone

In vivo and in vitro administration of the steroid molting hormone α-ecdysone or β-ecdysone (ecdysterone) to intact *Drosophila* intermolt larvae or isolated tissues, results in a definite change in the pattern of morphologically distinct puffs (Berendes, 1967). New puffs arise in *Drosophila hydei* in two separate phases: one group of three puffs develops within 15—20 min and a second group of five puffs appears 4—6 h after administration of the hormone. In addition to newly arising puffs, some of the puffs already present become more active, whereas other puffs undergo a reduction in their activity. Among these puffs, some may not be responding to the hormone but to a change in environmental factors (medium). The latter changes in the puffing pattern can also be observed after injection of the solvent (Ringer's solution) or after incubation of tissues in the solvent deprived of hormone (e.g. puff 97A in *D. hydei*).

The first changes in the puffing pattern become evident 15 min after injection of a hormone solution. The maximum size of newly arising puffs and of some of the puffs which increase in activity is attained between 40 and 50 min, if the optimal hormone concentration (10^{-2} mg/ml) is applied. With a range of suboptimal doses of the hormone (10^{-5}—10^{-2} mg/ml), a dosage-effect curve for the relationship between puff size and hormone concentration can be obtained (Leenders, Wullems and Berendes, 1970).

The changes in the puffing pattern following experimental application of the hormone are essentially the same as those occurring in normal development during the 6-h period prior to puparium formation. It is, however, not only the genome which responds similarly to an experimental and a natural raise in ecdysone titer.

2. Hormone-induced Changes in Salivary Gland Function

Experimental application of appropriate doses of molting hormone to mid third instar larvae of *D. hydei* affects certain metabolic features of the salivary gland cells, which occur in normal development only shortly before puparium formation. One of these features is the release from the cells of a glue substance which serves to attach the puparium to a substrate (Fraenkel, 1952; Fraenkel and Brookes, 1953; Perkowska, 1963). During most of the third larval instar, the glue substance, which has a mucopolysaccharide composition, is synthesized and stored as membrane-limited granules or droplets within the salivary gland cells. In *D. hydei* this product is suddenly released from these cells approximately one hour before puparium formation (BERENDES, 1965a). Recently, it was shown that the release of this cell product is controlled by the molting hormone. The secretion of mucopolysaccharide from the cells into the gland lumen can be induced in vivo as well as in vitro by providing the cells with molting hormone (POELS, 1970). Concomitant with the release of the mucopolysaccharide, a variety of changes occurs in the cytoplasmic organization of the salivary gland cells. Not only do the secretion granules, originally distributed randomly over the cytoplasm, become concentrated at the cell apex, but the distribution of mitochondria and the organization of the endoplasmatic reticulum are also altered in a specific manner. The same applies to the villi extending from the cell apex into the gland lumen: Their structure changes and there is a drastic increase in their number (POELS, DE LOOF, and BERENDES, 1971). These observations indicate that, among other cell types, the salivary gland cells should be regarded as target cells for the action of the molting hormone ecdysone.

3. Relationship between Hormone-induced Changes in Genome Activity and the Change in Cell Function

It seems obvious to raise, at this point, the question whether the changes in the pattern of genome activities induced by the hormone have some relationship with the evident structural reorganization at the cytoplasmic level involved in the hormone-controlled change in cell function. Two alternative possibilities for such a relationship should be considered. In the first instance, it could be assumed that the changes in the puffing pattern reflect an alteration in the pattern of information production which, at the cytoplasmic level, leads to a change in the functional capacities of the cell. On the other hand, it cannot be excluded that the activity of the genome is modified in response to a definite change in cytoplasmic properties brought about by the hormone. In addition to these alternatives, it could be supposed that the events at the different cellular levels, both induced by the hormone, are unrelated.

A first distinction between these alternatives could be made by studying the sequence in time of occurrence of the various phenomena following the experimental administration of hormone. Under optimal conditions with regard to the age of the larvae and the concentration of the steroid applied, it takes at least 4 h before secretion of the mucopolysaccharide can be observed. The first group of changes in the puffing pattern, however, occurs within the first hour. These changes in the activity of the genome could thus well be required for the changes at the cytoplasmic level. This suggestion was tested with isolated salivary glands maintained in an appropriate medium. Addition of ecdysone to the medium does induce hormone specific changes in genome activity as well as the secretion of mucopolysaccharides from the cells five to six hours later. Just as under in vivo conditions, the secretory event is accompanied by typical changes in cytoplasmic organization of the cells.

Inhibition of RNA synthesis or protein synthesis by actinomycin D or cycloheximide to a level below 5% of that of controls did inhibit mucopolysaccharide secretion from the cells. It should be pointed out that upon simultaneous addition of ecdysone and cycloheximide, the hormone specific puffs develop normally and reveal a strong incorporation of tritiated uridine. From these observations it may be inferred that the ultimate effect of hormone action in the salivary gland cells, the secretion of their mucopolysaccharide product, not only requires *de novo* RNA synthesis, but also *de novo* protein synthesis. On account of these data it is tempting to suppose that the RNA synthesized in the hormone specific puffs is an essential prerequisite for the changes in metabolism at the cytoplasmic level (Poels, 1972). Another observation, however, does not fit this idea. If puffing patterns of three functionally different tissues are compared in the same larva injected with the hormone, it becomes clear that a number of genome loci are affected in the same manner in all tissues (Berendes, 1966). In spite of this, the changes in submicroscopic structure of the different types of cell following their contact with the steroid hormone were definitely different (Berendes and Willart, 1971). The mechanism(s) by which cells with a dissimilar developmental history develop, in response to the molting hormone inducing a common response at the genome level, changes in structure and function typical for the type of cell, remains to be elucidated.

With respect to the second group of changes in the puffing pattern, it was established that these occur either concomitantly with or later than the onset of the secretory event (Poels, 1970).

4. Possible Mechanisms of Hormone Action

The mere fact that the molting hormone does affect the activity of certain genome loci should be useful in the search for the mechanism of hormone action. So far, two possible mechanisms have been proposed. A direct control of the activity of certain genes by an action of the hormone at the genome level was suggested by Karlson and his coworkers (Karlson, 1965; Karlson and Sekeris, 1966; Congote, Sekeris and Karlson, 1969). Kroeger, on the other hand, postulates the cell membrane as a primary site of hormone action. According to his hypothesis (Kroeger, 1963, 1966; Kroeger and Lezzi, 1966), changes in membrane permeability would give rise to an alteration of the intracellular ion balance with a consequent change in the activity of certain genes. A variety of experimental data having been provided as support for either one of these hypotheses, it seems plausible to consider the possibility that the two mechanisms may operate more or less simultaneously by affecting different genome loci in their activity. This suggestion seems to be supported by observations on the puffing pattern of salivary gland cells following incubation in a medium containing a non-physiological, high potassium concentration (Kroeger, 1963; Berendes, van Breugel and Holt, 1965). Although this treatment should influence the pattern of gene activities in the same way as a hormone treatment (Kroeger, 1963), only some of the loci in *D. hydei* known to respond to the hormone reveal a similar response to the KCl treatment. This indicates that only some of the changes in genome activity following hormone treatment may actually result from a change in the intracellular ion balance. If this interpretation is correct, it remains to be explained why the total pattern of changes in genome activity in response to a KCl treatment includes a number of loci which are not at all affected by the hormone! Anyhow, it appears that only a few loci among the group responding to hormone treatment may be affected indirectly by a change in the intracellular ion balance. Does this mean that the activity of the other loci is controlled by an interaction of the hormone at the genome level? If this were so, there should exist not only a mechanism for active

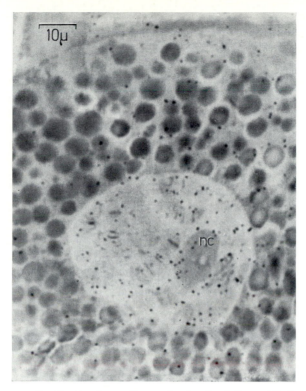

Fig. 4. ³H-ecdysone distribution in a salivary gland cell of *Drosophila hydei* approximately 8 hrs before puparium formation. Section of freeze-dried and OsO₄-fixed material. nc: nucleolus

transport of the hormone from the cell membrane to the inside of the nucleus, but also a device for recognition of these genome loci by the hormone.

It is well established that a variety of mammalian steroid hormones become associated with receptor proteins which probably serve as carriers for the intracellular hormone transport (Toft and Gorski, 1966; Jensen et al., 1968, 1969; Alberti and Sharp, 1969; Shyamala and Gorski, 1969; Milgrom and Beaulieu, 1970a, b; Musliner, Chader, and Villee, 1970; Schaumburg, 1970; O'Malley, Spelsberg, Schrader, Chytil, and Steggles, 1972). Recently, it was reported that the insect hormone, ecdysone, also appears to bind to proteins at the cytoplasmic as well as at the nuclear level (Emmerich, 1970, 1972). Moreover, autoradiographic studies on the intracellular distribution of the hormone have indicated that, in salivary gland cells of *D. hydei* and *D. virilis*, ecdysone becomes, at least at certain developmental stages, accumulated within the nuclei (Emmerich, 1969, Claycomb, Lafond, and Villee, 1971) (Fig. 4). But, despite the fact that ecdysone specific puffs were induced in these experiments, no indication was obtained of a specific accumulation of hormone molecules in these structures. However, because the specific activity of the ecdysone applied was rather low (2.9 Ci/mM), the failure to detect an evident accumulation does not necessarily exclude the possibility that the hormone actually arrives at the hormone-responsive loci and that a hormone-protein complex does interact with the local molecular configuration in such a manner that the local DNA can be transcribed. The finding of a certain proportion of hormone molecules associated with nonhistone chromatin-proteins may support the idea of direct control of genome activity by the steroid in insect target cells.

In contrast to the results described for *D. hydei*, autoradiographic studies on the intracellular localization of ecdysone in *Chironomus*, *Rhynchosciara* and *Calliphora* salivary gland cells failed to demonstrate, at any particular developmental stage investigated, a specific nuclear accumulation of the hormone (WEIRICH and KARLSON, 1969; THOMSON, ROGERS, GUNSON and HORN, 1970). Though these results do not offer an alternative explanation with regard to the mechanism of action of the hormone, they certainly fit the hypothesis of a primary action of the hormone at the cytoplasmic level. With regard to the obvious changes in cytoplasmic constituents induced by the hormone, it could be suggested that, in analogy to the mode of action of glucocorticoid hormones in cultured hepatoma cells (SAMUELS and TOMKINS, 1970), ecdysone interacts in a specific manner with receptor molecules regulating the synthesis of certain enzymes in the salivary gland cells. On the other hand, the experiments of POELS (1972) with isolated salivary glands suggested that genome activity is required for the specific hormone response at the cytoplasmic level, the secretion of mucopolysaccharides.

In contrast to the mechanism of puff induction activated by a rise in environmental temperature, in which the actual factor responsible for the initiation of specific genome activity is still uncertain, the hormone-controlled system offers the advantage that the effector is chemically well defined and can be applied in its pure form. This fact has increased our insight into the mode of action of the steroid at the cellular and subcellular levels. As yet, however, it still remains to be defined whether control of the activity of certain genes is exerted by a direct action of the steroid (or a steroid-protein complex) at the genome level rather than by secondary effectors activated by the hormone. It seems unlikely that the induction of hormone specific puffs results from a direct action of free steroid molecules at the genome level. Although in isolated nuclei incubated with the steroid the quantity of RNA synthesized within 20 min is increased as compared to control nuclei incubated without the hormone, competition hybridization experiments failed to reveal synthesis of new RNA species (ALONSO, 1972). Moreover, ecdysone did not induce the development of the specific puffs in isolated nuclei.

III. The Phenomenon of Puff Formation

Experimental application of temperature treatments and the hormone ecdysone have been used for the investigation of the process of puff formation in *Drosophila hydei*. Because detailed accounts of the information obtained from these studies with respect to the various factors involved in the mechanism of puff formation have been published recently (BERENDES, 1969, 1971), only the most pertinent facts will be summarized here.

The first indication for puff formation of a particular chromosome locus is the accumulation of non-histone protein in a restricted area of the presumptive puffing site. This accumulation can be demonstrated cytochemically by staining with a variety of dyes binding to proteins. In many instances, the accumulation of protein is visible as a sharply defined, colored band located in an interband region. Sometimes this band is clearly separated from any of the detectable chromosome bands, while in other regions it may be closely associated with a chromosome band. Following this initial phase, a chromosome band (or several bands) in the direct vicinity of the area of protein accumulation loses its condensed structure. Attempts to define the origin of a puff in terms of the decondensation of one particular band, which have been successful for a Balbiani ring in *Chironomus tentans* (BEERMANN, 1952, 1967), have so far failed in *Drosophila hydei*. Concurrently with the decondensation of band chromatin, the diameter of the chromosome region grows, the local non-histone protein concentration increases and the protein becomes incorporated over the entire puff region. The onset of RNA synthesis can be detected soon after the first protein accumulation has occurred.

By the time that the puffed region has attained its final size, the first characteristic 300—400 Å RNP-particles can be observed. The final size of the puffs seems to be quantitatively related to the stimulus applied. During the growing phase, the quantity of protein accumulated in the region appears to be roughly proportional to the size of the puff. After the ultimate size has been established, the protein quantity remains more or less constant or decreases very slowly (HOLT, 1970, 1971).

Autoradiographic analysis of the incorporation of tritiated uridine within active puffed regions indicates that RNA synthesis occurs, at least in a number of puffs, only at a definite site within the puff region. From its site of synthesis, the newly synthesized RNA migrates within the puff region either to one side from the area in which transcription occurs, or in both directions (BERENDES, 1969).

With respect to the chemical features, molecular size and base composition, and the metabolic function of the RNA synthesized in the individual, experimentally induced puffs of *Drosophila hydei*, nothing is known yet. However, the results of in situ hybridization studies with ribosomal RNA suggest that it is unlikely that these puffs are involved in the synthesis of ribosomal RNA (PARDUE, GERBI, ECKHARDT, and GALL, 1970).

A. Factors Involved in Puff Formation

1. The Initiation of Puff Formation

The data presented so far have all been obtained as a result of treatments applied to intact cells or even to the organism as a whole. It is evident that this approach is less adequate to resolve the mechanism of gene activation at the genome level. In order to acquire an insight into the process of initiation of transcription, it seems essential to raise primarily the question whether or not the genome response to a stimulus requires the intact system of nucleocytoplasmic interactions.

Isolated nuclei from a variety of cell types are known to retain some of their essential metabolic capacities, e.g. RNA synthesis and the incorporation of amino acids into TCA precipitable material, for a certain period following isolation (POGO, LITTAU, ALLFREY, and MIRSKEY, 1967; BURDMAN, JORNEY, and ROBINSON, 1968; TZUZUKI and NAORA, 1968; RISTOW and ARENDS, 1968; SCHMUCKLER and KOPLITZ, 1969; TZUZUKI, 1970; CESTARI and PAVAN, 1970; PIRRONE, MUNISTERI, ROCCHERI, MUTOLO, and GIUDICE, 1971). Also nuclei isolated from salivary gland cells of *Drosophila hydei* are capable of RNA synthesis and amino acid incorporation in vitro (BERENDES and BOYD, 1969; HELMSING, 1970). It appears that the chromosomal distribution of the incorporation of RNA precursors is identical to that observed under in vivo conditions.

Despite these functional capacities, isolated polytene nuclei fail to respond to stimuli like ecdysone or a temperature treatment with puff formation at the stimulus-specific loci. It is likely, however, that the mechanism active in puff growth is still intact in the isolated nuclei. This was indicated by a series of experiments in which isolated glands were incubated with ecdysone for very short periods prior to the isolation of nuclei. If intact cells are provided with ecdysone for at least 8 min, the ecdysone-specific puffs develop in the isolated nuclei within 40 min. After incubation of the glands for less than 8 min, no ecdysone puffs developed in the isolated nuclei. No ecdysone puffs were visible in squashes of the glands immediately after 8 min incorporation (Leenders, unpublished). It was assumed on account of these observations that in the absence of cytoplasm the mechanism of initiation of puff formation fails to function.

In order to test this assumption, isolated nuclei were incubated with cytosol fractions supplied with ecdysone. These experiments did not result in the initiation of hormone-specific puffs. On the other hand, there is indirect evidence which suggests that cytoplasmic factors may be required for puff induction.

2. Indications for Nucleo-cytoplasmic Interactions in Puff Formation

Since the newly induced puffs accumulate significant quantities of a non-histone type of protein, the pattern of SDS-phenol soluble (acidic) proteins extracted from nuclei isolated from cells subjected to an ecdysone or a temperature treatment was compared with that of nuclei isolated from untreated cells. The electropherograms of untreated cells were essentially the same, revealing 20 distinct bands. The nuclei of ecdysone-treated cells in which ecdysone puffs had developed revealed one clear band in addition to the basic pattern of untreated cells. The same band is present in acidic nuclear proteins extracted from salivary glands of animals shortly before puparium formation in which the ecdysone-specific puffs were present as a consequence of a natural increase in hormone titer at this stage. The molecular weight of this one extra protein band is close to 42000. A similar result was obtained with nuclei in which "temperature" puffs had been induced. Also in the electropherograms of acidic proteins extracted from these nuclei one definite band appeared which was absent in the controls (Fig. 5). This protein band was also present if the "temperature" puffs were induced by CO_2 treatment (see page 186). Comparison of the electropherograms of nuclei with ecdysone- and those with "temperature"-induced puffs clearly shows that the two novel protein bands differ in their location in the gels. The protein appearing in nuclei with temperature puffs has a molecular weight approximating 23000 (Helmsing and Berendes, 1971). It has still to be demonstrated that these new acidic protein fractions are derived from the newly induced puffs. On the other hand, it could be ascertained that the presence of these protein fractions in the nuclei is always correlated with the presence of the specific puffs irrespective of the type of nuclei studied. Apart from salivary gland nuclei, the protein fractions related with the presence of hormone- or temperature induced puffs could be isolated from midgut nuclei. Recently, it was shown that the 23000 MW protein fraction was present in chromatin prepared from salivary gland nuclei displaying temperature-specific puffs (Helmsing, 1972). The specific protein fractions isolated

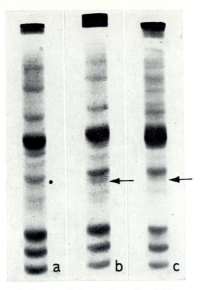

Fig. 5. Separation of sodium dodecyl sulphate-phenol soluble proteins extracted from mid third instar salivary gland nuclei of *Drosophila hydei* on 15% polyacrylamide gels. a control nuclei (untreated); b nuclei of larvae which had been submitted to a temperature treatment of 30 min prior to the isolation of the nuclei. The arrow indicates the presence of a novel protein fraction (approximate M. W. 23000) which is absent in the control nuclei a; c nuclei of larvae which had been submitted to a 45-min treatment with CO_2 followed by a 15-min period of recovery in fresh air. The arrow indicates the presence of a protein fraction at the same location as in b. The nuclei utilized in b and c displayed "temperature" puffs

from nuclei displaying experimentally induced puffs have one feature in common with the protein(s) accumulating within the puff loci during puff formation; their occurence is independent of de novo protein synthesis (HELMSING, 1972). It seems well established that these novel protein fractions originate from the cytoplasm.

There are further indications which favor the assumption that an intact nucleo-cytoplasmic interaction is essential for the initiation of puff activity. It was found that the response of the genome to ecdysone can be modified by affecting the adenyl cyclase activity. Cyclic AMP appears to enhance the effect of ecdysone as determined by measuring the size of ecdysone-induced puffs (LEENDERS, WULLEMS, and BEREN-DES, 1970).

Furthermore, ecdysone appears to become associated with proteins at the cyto-plasmic level and can also be extracted as hormone-protein complex from the nuclei (EMMERICH, 1970). It is unclear whether the complex at the nuclear level is derived from the cytoplasm. It was indicated, however, that pure hormone injected directly into the nucleus of a salivary gland cell is far less effective in producing hormone-specific puffs than hormone injected into the cytoplasm (VAN DER VELDEN, unpublish-ed). After hormone injection into the nucleus, the puffs take longer to appear and their size is significantly smaller than after injection of the same concentration into the cytoplasm. Although alternative explanations are possible, this result could be interpreted in terms of the absence of the appropriate protein receptors for hormone binding at the nuclear level.

In addition to factors at the cytoplasmic level which seem to be essential for the initiation of gene activity, factors of nuclear origin may be involved, too. It was indicated that some type of RNA molecules might act as derepressor molecules (Frenster, 1965). A similar function was attributed to certain RNA species in the recently proposed model for the regulation of gene activity in higher organisms (Britten and Davidson, 1969). It is not unlikely that some type of RNA is also involved in the process of puff induction in *Drosophila*. The injection of ribonuclease (10^{-7} μg/nucleus) directly into salivary gland nuclei appears to inhibit the response of these nuclei to a subsequent temperature treatment. It was ascertained that the failure of these nuclei to respond to the treatment was not merely a result of the injection procedure. Nuclei injected with the solvent gave a normal response by showing well-developed temperature puffs within 20 min after onset of a temperature treatment in vitro. Because it remained uncertain whether the inhibitory effect was caused by the enzyme activity or by the injection of the protein as such, oxidized ribonuclease in similar and higher concentrations was injected prior to a temperature treatment. Similarly, a protamine solution was injected. Neither the oxidized RNase, nor the protamine influenced the response of the nuclei to a temperature induction. It may thus be suggested that the enzymatic activity of the RNase is responsible for the observed inhibitory effect (Holt, van der Velden, and Berendes, unpublished).

A contradictory result is obtained if salivary glands are incubated with RNase (5 mg/ml) (see also Ritossa et al., 1965). In these experiments new puffs, in particular temperature puffs, develop after 4 to 5 h of incubation. However, this effect may be a consequence of a general disturbance of the energy metabolism of the cells occurring after long incubation times, which can also be observed in some glands incubated without RNase.

With regard to the effect of the RNase injected into nuclei, it could be supposed that it degrades some type of RNA involved in the initiation of puff formation. Since pretreatment of the salivary glands with high doses of actinomycin D (20 μg/ml) or α-amanitin (0.5 μg/larva) prior to a temperature treatment does not inhibit the formation of the specific puffs (Berendes, 1968; Holt and Kuypers, in press), this type of RNA should not result from a de-novo synthesis initiated by the treatment.

B. Quantitative Regulation of Puff Activity

It is well established that various degrees of suboptimal stimulation can produce various degrees of puffing, as expressed by puff size of the responding loci. Generally, the level of synthetic activity, estimated by autoradiographic analysis, is correlated with the size of a puff (Pelling, 1964). On account of these observations, it seems legitimate to suggest that the activity of a given puff, in terms of transcription, can vary within certain limits according to the strength of the stimulus applied. In other words, transcription may be quantitatively controlled. A similar conclusion may be drawn from studies on the phenomenon of dosage compensation (Mukherjee and Beermann, 1965; Mukherjee, 1966; Korge, 1970). In the polytene chromosome system, quantitative regulation of the activity of a locus (puff) could be brought about either by a variation in the number of chromatids transcribed in the region, or by a regulation of the number of RNA copies transcribed from each chromatid per unit of time. It may be that the variation in number of chromatids transcribed as a mechanism for quantitative control of gene activity can operate exclusively in polytene or polyploid systems. On the other hand, it would be expected that a mechanism for quantitative control of gene activity operating in diploid interphase nuclei could also function in polytenic interphase nuclei.

At the chromatid (haploid) level, quantitative regulation of transcription could be monitored by a variation in the frequency of initiation of the transcription of a *single* informational DNA sequence. In that case the number of RNA copies transcribed per unit of time could be, within a certain range, proportional to the availability (or concentration) of initiation factors, presuming that other factors required for transcription are sufficiently available. In such a model, maximum activity of the locus would be determined by the maximum number of RNA-polymerase molecules which can be simultaneously associated with the informational DNA sequence.

A quantitative control of the production of RNA copies from a definite DNA sequence could also be based on the presence of a serial repetition of the informational sequence at the chromatid level. Such an arrangement was recently proposed in a model for quantitative regulation of gene activity in higher organisms (GEORGIEV, 1969). In this model the number of RNA copies to be synthesized depends upon the number of successive identical DNA sequences subsequently transcribed by a RNA-polymerase molecule. Studies on the RNA metabolism at the puff level (Balbiani ring in *Chironomus tentans*) provided some data which could be interpreted according to this model (DANEHOLT et al., 1969; DANEHOLT, 1970) although alternative explanations cannot be excluded (PELLING, 1970). Definite evidence for the presence of redundant informational sequences in loci undergoing puffing has not yet been provided, however. On the other hand, it has been shown that in Diptera at least some functional genes are redundant. In *D. melanogaster* between 180 and 230 genes for the 18S and 28S ribosomal components (both derived from a common 37S precursor) are clustered in the nucleolus organizer region (RITOSSA, ATWOOD, and SPIEGELMAN, 1966). In the same species, the third ribosomal RNA component, 5S RNA, is presumably synthesized exclusively at a separate genome locus, which could be identified by in situ RNA-DNA hybridization as one band or a few bands in the polytene chromosome region 2—56F (WIMBER and STEFFENSEN, 1970). The 5S RNA genes, of which there are approximately 200 in the haploid genome, are also clustered (TARTOF and PERRY, 1970; QUINCEY, 1971). The presence of this number of repeated 5S RNA sequences within a restricted chromosome area, presumably one band, elucidates, at least for this area, the genetic significance of the excessive amount of DNA present at a morphologically defined chromosome unit. The occurrence of a certain sequence repetition in particular chromosome bands is also suggested by quantitative measurements of the DNA content of particular bands in the polytene chromosomes of the midge *Chironomus thummi thummi* in comparison to homologous bands in a closely related species, *Ch. th. piger* (KEYL, 1964, 1965, 1966). The presence of clusters of repeated sequences with a defined genetic function may provide a basis for quantitative regulation of the production of a definite RNA species, the level of production being controlled by variation of the number of copies simultaneously transcribed.

Whereas, in the model developed by GEORGIEV (1969), one initiation at the beginning of the series of identical sequences could eventually lead to the production of a large number of RNA copies of the same gene, quantitative regulation could equally well be based upon simultaneous initiation at various sites along a chain of redundant genes. Support for the latter mechanism has been obtained from an electron microscope study of the fine structure of active ribosomal genes in amphibian oocytes. In this special case, autonomous nucleolus organizers originating from the genome are composed of a series of genes active in the production of 40S ribosomal RNA precursor molecules (GALL, 1966). These genes are arranged successively along the nucleolus organizer DNA, separated from each other by a non-transcribed spacer region. Many of the ribosomal genes are transcribed simultaneously (MILLER and Beatty, 1969a, b). Apart from the simultaneous transcription of repeated sequences, each sequence can be transcribed by approximately one hundred polymerase molecules at the same time. Quantitative control of the number of 40S precursor molecules per unit of time could therefore be regulated not only by the number of sequences transcribed but also by the number of initiations of transcription per sequence.

So far, there is no evidence for the presence of large numbers of reiterated sequences in the puffs of *Drosophila hydei*, as far as can be concluded from autoradiographic studies on transcription in these regions. Large puffs occurring in the course of normal development, as well as induced by temperature treatments, reveal that after short pulses of tritiated uridine (1 min) the incorporation of this precursor is restricted to a definite area of the puffed region. An extension of the pulse period to 5 min reveals most of the label still present at the specific area, though label is found at both sides of this area (puff 81B) (BERENDES, 1969). In regarding puff 81B, induced by a temperature treatment, as a model, it may be concluded that in this puff only a definite area of the extended DNA is actually transcribed. Electron microscopic observations of this region revealed extended chromatin fibers arranged more or less parallel over the *entire* puff region. If these fibers are composed of a number of successive repeated sequences, these are certainly not transcribed simultaneously. Simultaneous transcription of the successive repetitions would produce, even after a 1-min pulse, overall labeling of the puff region. An interpretation of the observed label distribution in terms of a snapshot at some moment during the transcription of a series of repeated sequences is not supported by the fact that, in all cases observed, the label occurs at the same location within the puffed locus. Furthermore, the transverse array of grains over the puff region indicates simultaneous transcription of the same site at different homologous chromatids. This aspect alone would require a special regulatory device.

It could be supposed that the transcription of the extended DNA in the puff region, if it were composed of repeated sequences, would occur according to a master-slave concept, similar to that proposed for transcription of the extended DNA in the loops of lampbrush chromosomes (CALLAN and LOYD, 1960; CALLAN, 1967). This suggestion seems to be favored by the observation that, in some puffs in which transcription is restricted to a particular site, label is distributed from this site unidirectionally over the puffed region (BERENDES, 1968). Moreover, typical ribonucleoprotein particles, originating within the puff region, are generally found to be associated with the extended nucleohistone fibrils over their entire length in the puff.

In at least one puff, 2—48C (Fig. 1), it appears that the RNP particles migrate uni-directionally within the puff from their site of origin to an area where aggregation occurs. Within puff 2—48C, the direction of migration of the RNP particles is identical with that of uridine-labeled material from the site of transcription (see BERENDES, 1968).

In other puffs, however, it appears that newly synthesized RNA migrates from the site of transcription in both, opposite, directions over the puff region. So far, puff 2—48C is the only puff in which aggregation of ribonucleoprotein particles has been observed, thus no further indications are available as to the direction of migration of newly synthesized RNA within the puff region.

Studies on the submicroscopic organization of the transcription site within experimentally induced puffs in *D. hydei* failed to produce evidence for any kind of looped structure, as would be expected if slaves had to match with a master sequence before being transcribed (CALLAN, 1967). It thus seems unlikely that transcription in the puffs of *D. hydei* occurs according to the master-slave concept, despite the indications for a unidirectional transition of the product of transcription in puff 2—48C.

From the data so far presented, it may be concluded that transcription in various puffs in *D. hydei* is restricted to a definite site of the chromatid DNA extended in the puffed region. If this site consisted of only one single transcription unit, producing a unique type of informational or regulatory RNA, quantitative control of the production of this RNA species could be monitored solely by variation of the number of initiations of transcription per unit of time. If this were so, the frequency of initiation could, within certain limits (the maximum number of RNA polymerase molecules on the informational DNA sequence being determined by spatial and or structural parameters), be related to the quantitative availability of initiator molecules or complexes. Exhaustion of the initiator would necessarily lead to a cessation of transcription. This is what is actually observed if a stimulus is removed. When larvae subjected to a temperature treatment at 37° C are replaced at 25°, transcription in the temperature puffs ceases and the puffs regress. This result indicates that, in addition to the turning off of the supply of initiator molecules or complexes as a consequence of the withdrawal of the stimulus, another mechanism must be active by which initiator molecules or complexes available from previous stimulation can be disintegrated or inactivated. It could be assumed that inactivation of the initiator is a consequence of the initiating event. However, this mechanism could not account for all instances in which a site active in transcription ceases its activity.

In *Chironomus tentans*, a particular ecdysone-specific puff (CLEVER, 1966) regresses after a certain period of activity in spite of the fact that the level of hormone in the hemolymph is sufficient to induce the activity of the same puff at a younger stage. Similary, ecdysone puffs in *D. melanogaster* induced *in vitro*, regress after a certain period of activity in spite of the presence of sufficient hormone in the medium (ASHBURNER, 1971). In *D. hydei*, some ecdysone puffs fail to respond to a hormone injection if they have just finished a period of activity (BERENDES, 1967). A similar phenomenon has been described for temperature puffs which fail to respond to the stimulus after it has been applied repeatedly in alternation with periods of recovery at 25° C (BERENDES, 1969).

The failure of a puff to respond to its specific and adequate stimulus and the regression of a puff in the presence of an adequate level of its stimulus may have a common origin. It is tempting to assume that, in these instances, the formation of initiator molecules or complexes at the cytoplasmic level is inhibited.

C. Genetic Aspects of Puff-forming Ability

For numerous loci displaying a puffed structure some time during late larval development, the ability to form puffs appears to be independent of the position of the locus characterized by the gene sequence in which it is incorporated. A change in this position through chromosomal rearrangement generally does not affect the puff-forming capacity. This conclusion is supported by the fact that comparisons of puffing patterns in homologous chromosome sections of heterozygous rearrangements reveal no distinct differences, neither in cases in which the homologous sections are synapsed, nor in those in which the homologues are asynapsed. The relative autonomy of the puff-forming ability of certain loci was further demonstrated by the use of temperature treatments as a tool for investigating the response of sensitive loci after their transposition by chromosomal rearrangement. It was found that the response of two temperature-sensitive loci in *D. melanogaster*, 87A and 87B respectively, is identical, irrespective of a change in their location by translocations. This indicates that the physical gene sequence in which a locus is incorporated may not be critical for the expression of its puff-forming behavior in response to the appropriate stimulus (Ellgaard and Brosseau, 1969). A similar conclusion may be drawn from a study of the activation of temperature-sensitive loci in a number of closely related species. A number of species of the repleta group of the genus *Drosophila* was used, i.e. *D. repleta, D. hydei, D. eohydei, D. neohydei, D. nigrohydei, D. mercatorum, D. buzzatti, D. mulleri, D. hamatofila* and *D. bifurca*. These species, belonging to different subgroups of the repleta group, may have originated from a common ancestor by the occurrence of numerous chromosomal rearrangements, in particular, paracentric inversions (Patterson and Stone, 1952; Wasserman, 1960). All species revealed in response to a temperature treatment the occurrence of puff formation at four loci which could, by analysis of the banding pattern, be defined as identical loci. Some of these loci had entirely different locations within the chromosomes as a consequence of evolutionary changes in gene sequence (Berendes, 1965b).

It thus appears that the four loci responding to a temperature treatment have retained their puff-forming ability in spite of drastic changes in gene sequences during species evolution.

There are, however, exceptions to the rule that transposition of a locus should not affect its puff-forming ability. If puff-forming loci are transmitted to a position adjacent to heterochromatin, in particular to centric heterochromatin, inhibition of puff formation may occur (Hartmann-Goldstein, 1966; Schultz, 1965; Mukherjee and Gupta, 1966). A particularly interesting observation was that of a differential inhibition of puff formation in asynapsed homologues, one of which is physically connected with heterochromatin, whereas the other is not. On the homologue attached to the heterochromatin, no puff formed at a locus which was clearly puffed in the homologue disconnected from the heterochromatin (Rudkin, 1963; Mukherjee and Gupta, 1966). As already pointed out by Schultz (1965), the inhibitory effect of heterochromatin on the activity of adjacent genes may be due to compaction of chromosome regions in the vicinity of heterochromatin. This might result in a mechanical inability to form puffs. The molecular basis of this heterochromatin position effect is not yet understood.

A change in puff-formation ability may also result from mutation, as indicated by the occurrence of heterozygous puffs in a variety of species (Hsu and Liu, 1948; Beermann, 1961; Panitz, 1965; Pavan and Perondini, 1967; Perondini and Dessen, 1969; Ashburner, 1967b, 1969b). Heterozygous puffs have been interpreted as manifestations of a mutation in the structural gene of the puff-forming locus in one of the homologues (Ashburner, 1969). This interpretation may also be applied to the behavior of puff 64C in hybrids of *D. melanogaster* Oregon and a selection line vg6. Puff 64C becomes active at a late stage of larval development in vg6 larvae. It was never observed in the Oregon stock. In hybrids displaying asynapsed homologues including region 64C, a puff is found in one homologue only. These observations indicate that the ability for puff formation at locus 64C is inherited as a mendelian character. However, it appears that in synapsed homologues puff 64C is, though always smaller than in the homozygous parental strain, present on both homologues, and not as a heterozygous puff, as would be expected (Ashburner, 1967b). The question whether or not, in this particular example, and possibly in other cases described by Ashburner (1969b, 1970c), the locus on the homologue which in an asynapsed situation fails to show puff formation does so because of a mutation in a regulatory part of the gene, has to be studied further. It may be concluded that studies on the behavior of heterozygous puffs in general could provide essential data for a better understanding of the genetic fine structure underlying the functional organization of puff-forming loci.

IV. Conclusions

At first sight it may seem a little redundant to summarize the results so far obtained from studies on experimentally induced puffs so soon after the appearance of several reviews on the puffing phenomenon in the literature (Ashburner, 1970b, Berendes, 1971). It may be worthwhile, however, to attempt to give as a conclusion to the present discussions some outlook on future goals which may be reached as a result of further investigations of experimentally induced puffs. It should be pointed out that the changes in the basic puffing pattern originating from experimental treatments like temperature changes and hormone injections, are overall reproductions of the events occurring in normal development. It is this aspect which favors the use of these agents for studies on the mechanism of control of gene activity in a higher organism.

In the first instance, the control of the puffing phenomenon itself should receive further attention. Though some of the factors involved in the puffing process have been identified, their functional relationships are far from completely understood. The phenomenon of puff regression, if investigated in more detail, may provide essential information with regard to the control of transcription. It seems that, in particular, those puffs which regress in the presence of a sufficient quantity of the factor responsible for the induction of their activity are a suitable tool for the study of gene repression (Clever and Romball, 1966).

At the biochemical level, a characterization of the final product of certain puffs may be one of the primary aims. Though definite progress has been made on the isolation and characterization of puff RNA (Daneholt et al., 1969; Daneholt, 1970; Pelling, 1972), the final product of a puff, the ribonucleoprotein particles, are still to be characterized. It may be relevant to recall in this context the presence of characteristic aggregates of RNP particles in certain puffs in *Drosophila hydei* and *D. virilis* (Swift, 1964). The fact that these aggregates are present in only one particular puff may offer an opportunity for isolation of the final product of this puff. In *Drosophila hydei*, this puff can be experimentally induced by temperature treatment. The isolation of this typical puff product in sufficient quantity could be used to characterize the functional properties of the RNA species synthesized by a single puffing locus. The available methods for mass isolation of tissue and polytene nuclei from them (Boyd, Berendes and Boyd, 1968; Berendes and Boyd, 1969; Cohen and Gotchel, 1969; 1971) may provide the basis for the development of a technique for isolation of RNP particles from the nuclei. Such approach could give some insight into the significance of the synthetic activity of certain chromosome regions for cellular metabolism. So far, the significance of changes in the puffing pattern in relation to cellular activity has had to be deduced from correlated events at different cellular levels (Beermann, 1961; Grossbach, 1969; Poels, 1970).

Experimental puff induction may also be applied to the study of the specific relationship between inducing agent and responding gene loci. It seems reasonable to suggest that a continued search for the primary site of action of the temperature treatment, as well as the hormone, may in the future reveal an actual "trigger" for activation of a chromosome locus. A characterization of specific non histone proteins migrating from the cytoplasm into the nucleus during and following experimental induction of a particular group of genome loci, may illuminate one of the features of the relationship between an agent and its specific derepressing effect at the genome level.

In the past, the analysis of the puffing phenomenon has provided one of the most elegant demonstrations of the occurrence of differential gene activity in differentiated cells of a higher organism.

Is it unreasonable to suppose that the same phenomenon will in the future also reveal essential features of the mechanisms controlling the integrated machinery of differentiated cells of a eukaryote?

Acknowledgement

The author is much indebted to Dr. H. J. Leenders for providing him with unpublished results on the induction and repression of temperature puffs as recorded in sections II. A. 2 and A. 3.

References

Alberti, K.G.M.M., Sharp, G.W.G.: Macromolecular binding of aldosterone in the toad bladder. Biochim. biophys. Acta (Amst.) **192**, 335—346 (1969).

Alfert, M., Geschwind, I.I.: A selective staining method for the basic proteins of cell nuclei. Proc. nat. Acad. Sci. (Wash.) **39**, 991—999 (1953).

Alonso, C.: The effects of gibberellic acid upon developmental processes in *Drosophila hydei*. Entomol. exp. Appl. **14**, 73—82 (1971).

— The influence of molting hormone on RNA synthesis in isolated polytene nuclei of *Drosophila*. Develop. Biol. (in press).

ASHBURNER, M.: Patterns of puffing activity in the salivary glands of *Drosophila* I. Autosomal puffing patterns in a laboratory stock of *Drosophila melanogaster*. Chromosoma (Berl.) **21**, 398—428 (1967a).
— Gene activity dependent on chromosome synapsis in the polytene chromosomes of *Drosophila melanogaster*. Nature (Lond.) **214**, 1159—1160 (1967b).
— IV. Variability of puffing patterns. Chromosoma (Berl.) **27**, 156—177 (1969a).
— The genetic control of puffing in polytene chromosomes. In: Chromosomes Today, Vol. 2. Edinburgh: Oliver and Boyd 1969b.
— V. Responses to environmental treatments. Chromosoma (Berl.) **31**, 356—376 (1970a).
— Function and structure of polytene chromosomes during insect development. Advanc. Insect Physiol. **7**, 1—95 (1970b).
— The genetic analysis of puffing in polytene chromosomes of *Drosophila*. Proc. roy. Soc. Lond. B. **176**, 319—327 (1970c).
— Induction of puffs in polytene chromosomes of in vitro cultured salivary glands of *Drosophila melanogaster* by ecdysone and ecdysone analogues. Nature New Biol. **230**, 222—223 (1971).
BAUDISCH, W., PANITZ, R.: Kontrolle eines biochemischen Merkmals in den Speicheldrüsen von *Acricotopus lucidus* durch einen Balbianiring. Exp. Cell Res. **49**, 470—476 (1968).
BEERMANN, W.: Chromomerenkonstanz und spezifische Modifikationen der Chromosomenstruktur in der Entwicklung und Organdifferenzierung von *Chironomus tentans*. Chromosoma (Berl.) **5**, 139—198 (1952).
— Ein Balbianiring als Locus einer Speicheldrüsenmutation. Chromosoma (Berl.) **12**, 1—25 (1961).
— Gene action at the level of the chromosome. In: BRINK, R. A. (Ed.): Heritage from Mendel, pp. 179—201. Univ. Wisconsin Press 1967.
BECKER, H. J.: Die Puffs der Speicheldrüsenchromosomen von *Drosophila* melanogaster. I. Mitt. Beobachtungen zum Verhalten des Puffmusters im Normalstamm und bei zwei Mutanten, giant and lethal-giant larvae. Chromosoma (Berl.) **10**, 654—678 (1959).
— II. Mitt. Die Auslösung der Puffbildung, ihre Spezifität und ihre Beziehung zur Funktion der Ringdrüse. Chromosoma (Berl.) **13**, 341—384 (1962).
BERENDES, H. D.: Salivary gland function and chromosomal puffing patterns in *Drosophila hydei*. Chromosoma (Berl.) **17**, 35—77 (1965a).
— Gene homologies in different species of the repleta group of the genus *Drosophila*. I: Gene activity after temperature shocks. Genen en Phaenen **10**, 32—41 (1965b).
— Gene activities in the Malpighian tubules of *Drosophila hydei* at different developmental stages. J. exp. Zool. **162**, 209—218 (1966).
— The hormone ecdysone as effector of specific changes in the pattern of gene activities of *Drosophila hydei*. Chromosoma (Berl.) **22**, 274—293 (1967).
— Factors involved in the expression of gene activity. Chromosoma (Berl.) **24**, 418—437 (1968).
— Activités des chromosomes polyteniques et des chromosomes plumeux. Induction and control of puffing. Ann. Embryol. Morphogen. Suppl. **1**, 153—164 (1969).
— Gene activation in Dipteran polytene chromosomes. In: SEB Symposium 25. Cambridge: Univ. Press, pp. 145—161 (1971).
— BOYD, J.B.: Structural and functional properties of polytene nuclei isolated from salivary glands of *Drosophila hydei*. J. Cell Biol. **41**, 591—599 (1969).
— BREUGEL, F. M. A. VAN, HOLT, TH.K.H.: Experimental puffs in *Drosophila hydei* salivary gland chromosomes. Chromosoma (Berl.) **16**, 35—47 (1965).
— THIJSSEN, W. T. M.: Developmental changes in genome activity in *Drosophila lebanonensis casteeli* Pipkin. Chromosoma (Berl.) **33**, 345—360 (1971).
— WILLART, E.: Ecdysone related changes at the nuclear and cytoplasmic level of Malpighian tubule cells in *Drosophila*. J. Insect Physiol. **17**, 2337—2350 (1971).
BOYD, J.B., BERENDES, H.D., BOYD, H.: Mass preparation of nuclei from the larval salivary glands of *Drosophila hydei*. J. Cell Biol. **38**, 369—376 (1968).
BREUGEL, F.M.A. VAN: Puff induction in larval salivary gland chromosomes of *Drosophila hydei* Sturtevant. Genetica **37**, 17—28 (1966).

Britten, R. J., Davidson, E. H.: Gene regulation for higher cells: A theory. Science **165**, 349—357 (1969).

Burdette, W. J., Anderson, R.: Conditioned response of salivary gland chromosomes of *Drosophila melanogaster* to ecdysones. Genetics **51**, 625—633 (1965).

Burdman, J. A., Jorney, L. J., Robinson, L.: Protein synthesis in isolated nuclei from rat brain. J. Cell Biol. **39**, 164a (1968).

Callan, H. G.: The organization of genetic units in chromosomes. J. Cell Sci. **2**, 1—7 (1967).

— Lloyd, L.: Lampbrush chromosomes. In: New approaches in Cell Biology, pp. 23—46. New York: Academic Press 1960.

Cestari, A. N., Pavan, C.: DNA, RNA and protein synthesis in isolated nuclei and chromosomes of *Rhynchosciara*. J. Cell Biol. **47**, 32a (1970).

Claycomb, W. C., LaFond, R. E., Villee, C. A.: Autoradiographic localization of ^{3}H-β-ecdysone in salivary gland cells of *Drosophila virilis*. Nature (Lond.) **234**, 302—304 (1971).

Clever, U.: Induction and repression of a puff in *Chironomus tentans*. Develop. Biol. **14**, 421—438 (1966).

— Karlson, P.: Induktion von Puffveränderungen in den Speicheldrüsenchromosomen von *Chironomus tentans* durch Ecdyson. Exp. Cell Res. **20**, 623—627 (1960).

— Romball, C. G.: RNA and protein synthesis in the cellular response to a hormone, ecdysone. Proc. nat. Acad. Sci. (Wash.) **56**, 1470—1476 (1966).

Cohen, L. H., Gotchel, B. V.: Histones and other proteins in *Drosophila* polytene nuclei. Fed. Proc. **28**, 800 (1969).

— Histones of polytene and non-polytene nuclei of *Drosophila melanogaster*. J. biol. Chem. **246**, 1841—1848 (1971).

Congote, L. F., Sekeris, C. E., Karlson, P.: On the mechanism of hormone action. XIII. Stimulating effects of ecdysone, juvenile hormone and ions on RNA synthesis in fat body cell nuclei from *Calliphora erythrocephala* isolated by a filtration technique. Exp. Cell Res. **56**, 338—346 (1969).

Crouse, H. V.: The role of ecdysone in DNA-puff formation and DNA synthesis in the polytene chromosomes of *Sciara coprophila*. Proc. nat. Acad. Sci. (Wash.) **61**, 971—978 (1968).

Daneholt, B.: Base ratios in RNA molecules of different sizes from a Balbianiring. J. molec. Biol. **49**, 381—391 (1970).

— Edström, J.-E., Egyhazi, E., Lambert, B., Ringborg, U.: RNA synthesis in a Balbiani-ring in *Chironomus tentans* salivary gland cells. Chromosoma (Berl.) **28**, 418—429 (1969).

Ellgaard, E. G., Brosseau, G. E.: Puff forming ability as a function of chromosomal position in *Drosophila melanogaster*. Genetics **62**, 337—341 (1969).

Emmerich, H.: Anreicherung von Tritiummarkiertem Ecdyson in den Zellkernen der Speicheldrüsen von *Drosophila hydei*. Exp. Cell Res. **58**, 261—270 (1969).

— Ecdysonbindende Proteinfaktoren in den Speicheldrüsen von *Drosophila hydei*. Z. vergl. Physiol. **68**, 385—402 (1970).

— Ecdysone binding proteins in nuclei and chromatin from *Drosophila* salivary glands. J. gen. comp. Endocrin. (in press) (1972).

Fedoroff, N., Milkman, R.: Specific puff induction by tryptophan in *Drosophila* salivary chromosomes. Biol. Bull. **127**, 369 (1964).

Fraenkel, G.: A function of the salivary glands of the larvae of *Drosophila* and other flies. Biol. Bull. **103**, 285—286 (1952).

— Brookes, V. J.: The process by which the puparia of many species of flies become fixed to a substrate. Biol. Bull. **105**, 442—449 (1953).

Frenster, J. H.: A model of specific de-repression within interphase chromatin. Nature (Lond.) **206**, 1269—1270 (1965).

Gall, J. G.: Nuclear RNA of the salamander oocyte. Nat. Cancer Inst. Monogr. **23**, 475—488 (1966).

Georgiev, G. P.: On the structural organization of operon and the regulation of RNA synthesis in animal cells. J. theor. Biol. **25**, 473—490 (1969).

Grossbach, U.: Chromosomen-Aktivität und biochemische Zelldifferenzierung in den Speicheldrüsen von *Camptochironomus*. Chromosoma (Berl.) **28**, 136—187 (1969).

HARTMANN-GOLDSTEIN, I. J.: Relationship of heterochromatin to puffs in a salivary gland chromosomes of *Drosophila*. Naturwissenschaften **53**, 91 (1966).

HELMSING, P. J.: Protein synthesis of polytene nuclei in vitro. Biochim. biophys. Acta (Amst.) **224**, 579—587 (1970).

— Induced accumulation of nonhistone proteins in polytene nuclei of *Drosophila* II. Accumulation of proteins in polytene nuclei and chromatin of different larval tissues. Cell Differentiation 1, (in press) (1972).

— BERENDES, H. D.: Induced accumulation of nonhistone proteins in polytene nuclei of *Drosophila* hydei. J. Cell Biol. **50**, 893—896 (1971).

HOLT, TH. K. H.: Local protein accumulation during gene activation. I. Quantitative measurements on dye binding capacity at subsequent stages of puff formation in *Drosophila hydei*. Chromosoma (Berl.) **32**, 64—78 (1970).

— II. Interferometric measurements of the amount of solid material in temperature induced puffs of *Drosophila hydei*. Chromosoma (Berl.) **32**, 428—435 (1971).

— KUYPERS, A. M. C.: Induction of chromosome puffs in *Drosophila hydei* salivary glands after inhibition of RNA synthesis by α-amanitin. (Submitted for publication).

HSU, T. C., LIU, T. T.: Microgeographic analysis of chromosomal variation in a chinese species of *Chironomus* (Diptera). Evolution **2**, 49—57 (1948).

JENSEN, E. V., SUZUKI, T., KAWASHIMA, T., STUMPF, W. E., JUNGBLUT, P. W., DESOMBRE, E. R.: A two-step mechanism for the interaction of estradiol with rat uterus. Proc. nat. Acad. Sci. (Wash.) **59**, 632—638 (1968).

— NUMATA, M., SMITH, S., SUZUKI, T., BRECHER, P. I., DESOMBRE, E. R.: Estrogen-receptor interactions in target tissues. Develop. Biol. Suppl. **3**, 151—171 (1969).

KARLSON, P.: Biochemical studies of ecdysone control of chromosomal activity. J. cell. comp. Physiol. **66**, 69—76 (1965).

— SEKERIS, C. E.: Ecdysone, an insect steroid hormone, and its mode of action. In: PINCUS, G. (Ed.): Recent Progress in Hormone Research, pp. 473—502. New York-London: Academic Press 1966.

KEYL, H.-G.: Verdopplung des DNS-Gehalts kleiner Chromosomenabschnitte als Faktor der Evolution. Naturwissenschaften **51**, 46—47 (1964).

— Duplikationen von Untereinheiten der chromosomalen DNS während der Evolution von *Chironomus thummi*. Chromosoma (Berl.) **17**, 139—180 (1965).

— Probleme der biologischen Reduplikation. In: Funktionelle und morphologische Organization der Zelle, S. 55—69. Berlin-Heidelberg-New York: Springer 1966.

KIKNADZE, I. I.: Functional changes of giant chromosome under conditions of inhibited RNA synthesis. Tsitologia **7**, 311—318 (1965).

KORGE, G.: Dosiskompensation und Dosiseffekt für RNS-synthese in Chromosomen-Puffs von *Drosophila melanogaster*. Chromosoma (Berl.) **30**, 430—464 (1970).

KROEGER, H.: The induction of new puffing patterns by transplantation. Chromosoma (Berl.) **11**, 129—145 (1960).

— Chemical nature of the system controlling gene activities in insect cells. Nature (Lond.) **200**, 1234—1235 (1963).

— Zellphysiologische Mechanismen bei der Regulation von Genaktivitäten in den Riesenchromosomen von *Chironomus thummi*. Chromosoma (Berl.) **15**, 36—70 (1964).

— Potentialdifferenz und Puffmuster: Electrophysiologische und cytologische Untersuchungen an den Speicheldrüsen von *Chironomus thummi*. Exp. Cell Res. **41**, 64—80 (1966).

— Gene activities during insect metamorphosis and their control by hormones. In: ETKIN, W., GILBERT, L. I. (Eds.): Metamorphosis: a problem in developmental biology. New York: Appleton Century Crofts 1968.

— LEZZI, M.: Regulation of gene action in insect development. Ann. Rev. Entomol. **11** 1—22 (1966).

LAUFER, H., HOLT, TH. K. H.: Juvenile hormone effects on chromosomal puffing and development in *Chironomus thummi*. J. exp. Zool. **173**, 341—352 (1970).

LEENDERS, H. J.: The effect of changes in the respiratory metabolism upon genome activity in *Drosophila*. I. The induction of gene activity. (in press) (1972).

— BECKERS, P. J. A.: II. Feed back control of induced gene activity. (in press) (1972).

Leenders, H. J., Wullems, G. J., Berendes, H. D.: Competitive interaction of adenosine 3',5'-monophosphate on gene activation by ecdysterone. Exp. Cell Res. 63, 159—164 (1970).

Lezzi, M., Gilbert, L. I.: Control of gene activities in the polytene chromosomes of *Chironomus tentans* by ecdysone and juvenile hormone. Proc. nat. Acad. Sci. (Wash.) 64, 498—503 (1969).

Mahler, H. R., Cordes, E. H.: Biological Chemistry. New York-Evanston-London; Harper and Row 1969.

Milgrom, E. Beaulieu, E.-E.: Progesterone in the uterus and the plasma. I. Binding in rat uterus 105 000 g supernatant. Endocrinology 87, 276—287 (1970a).

— II. The role of hormone availability and metabolism om selective binding to uterus protein. Biochim. Biophys. Res. Commun. 40, 723—730 (1970b).

Miller, O. L., Beatty, B. R.: Visualization of nucleolar genes. Science 164, 955—957 1969a).

— Extrachromosomal nucleolar genes in amphibian oocytes. Genetics Suppl. 61, 134—143 (1969b).

Mukherjee, A. S.: Dosage compensation in *Drosophila*: an autoradiographic study. Nucleus 9, 83—96 (1966).

— Effect of dicyandiamide on puffing activity and morphology of salivary gland chromosomes of *Drosophila melanogaster*. Ind. J. exp. Biol. 6, 49—51 (1968).

— Beermann, W.: Synthesis of ribonucleic acid by the X chromosomes of *Drosophila melanogaster* and the problem of dosage compensation. Nature (Lond.) 207, 785—786 (1965).

— Gupta, A. D.: The role of heterochromatin in the control of gene activity. Dros. Inf. Serv. 41, 160—161 (1966).

Musliner, T. A., Chader, G. J., Villee, C. A.: Studies on estradiol receptors of the rat uterus. Nuclear uptake in vitro. Biochemistry 9, 4448—4453 (1970).

O'Malley, B. W., Spelsberg, T. C., Schrader, W. T., Chytil, F., Steggles, A. W.: Mechanisms of interaction of a hormone-receptor complex with the genome of a eukaryotic target cell. Nature (Lond.) 235, 141—144 (1972).

Panitz, R.: Heterozygote Funktionsstrukturen in den Riesenchromosomen von *Acricotopus lucidus*. Chromosoma (Berl.) 17, 199—218 (1965).

— Funktionelle Veränderungen an Riesenchromosomen nach Behandlung mit Gibberellinen. Biol. Bull. 86, Suppl. 147—156 (1968).

Pardue, M. L., Gerbi, S. A., Eckhardt, R. A., Gall, J. G.: Cytological localization of DNA complementary to ribosomal RNA in polytene chromosomes of *Diptera*. Chromosoma (Berl.) 29, 268—290 (1970).

Patterson, J. T., Stone, W. S.: Evolution in the Genus *Drosophila*. New York: Macmillan Comp. 1952.

Pavan, C., Perondini, A. L. P.: Heterozygous puffs and bands in *Sciara ocellaris* Domstock (1882). Exp. Cell Res. 48, 202—205 (1967).

Pelling, C.: Ribonukleinsäure-Synthese der Riesenchromosomen. Autoradiographische Untersuchungen an *Chironomus tentans*. Chromosoma (Berl.) 15, 71—122 (1964).

— Puff RNA in polytene chromosomes. Cold Spring Harbor Symp. Quant. Biol. 35, 521—531 (1970).

Perkowska, E.: Some characteristics of the salivary gland secretion of *Drosophila virilis*. Exp. Cell Res. 32, 259—271 (1963).

Perondini, A. L. P., Dessen, E. M.: Heterozygous puffs in *Sciara ocellaris*. Genetics Suppl. 61, 251—260 (1969).

Pirrone, A. M., Munisteri, A., Roccheri, M., Mutolo, V., Giudice, G.: Synthesis of RNA in isolated nuclei of sea urchin embryos. Wilh. Roux Arch. Entwickl.-Mech. Org. 167, 83—88 (1971).

Poels, C. L. M.: Time sequence in the expression of various developmental characters induced by ecdysterone in *Drosophila hydei*. Develop. Biol. 23, 210—225 (1970).

— deLoof, A., Berendes, H. D.: Functional and structural changes in *Drosophila* salivary gland cells as a consequence of ecdysterone. J. Insect Physiol. 17, 1717—1729 (1971).

Pogo, A. O., Littau, V. C., Allfrey, V. G., Mirsky, A. E.: Modification of ribonucleic acid synthesis in nuclei isolated from normal and regenerating liver: Some effects of salt and specific divalent cations. Proc. nat. Acad. Sci. (Wash.) 57, 743—750 (1967).

QUINCEY, R. V.: The number and location of genes for 5S ribonucleic acid within the genome of *Drosophila melanogaster*. Biochem. J. **123**, 227—233 (1971).

RISTOW, H., ARENDS, S.: A system in vitro for the synthesis of RNA and protein by isolated salivary glands and by nuclei from *Chironomus* larvae. Biochim. biophys. Acta (Amst.) **175**, 178—186 (1968).

RITOSSA, F. M.: A new puffing pattern induced by temperature shock and DNP in *Drosophila*. Experientia **18**, 571—572 (1962).

— Experimental activation of specific loci in polytene chromosomes of *Drosophila*. Exp. Cell Res. **35**, 601—607 (1964).

— ATWOOD, K. C., SPIEGELMAN, S.: On the redundancy of DNA complementary to amino acid transfer RNA and its absence from the nucleolar organizer region of *Drosophila melanogaster*. Genetics **54**, 663—676 (1966).

— PULIZER, J. F., SWIFT, H., BORSTEL, R. C. VON: On the action of ribonuclease in salivary gland cells of *Drosophila*. Chromosoma (Berl.) **16**, 144—151 (1965).

RUDKIN, G. T.: The structure and function of heterochromatin. In: GEERTS, S. J. (Ed.): Genetics Today **2**, 359—374 (1963).

SAKHAROVA, M. N., RAPOPORT, I. A., BEKNAZAR'YANTS, M. M., NIKIFOROV, Y. L.: Puffs induced by thiocyanate and a puff model for determining drugaffected enzymes. Doklady Akad. Nauk. SSSR **196**, 1217—1220 (1971).

SAMUELS, H. H., TOMKINS, G. M.: Relation of steroid structure to enzyme induction in hepatoma tissue culture cells. J. molec. Biol. **52**, 57—74 (1970).

SCHAUMBURG, B. P. S.: Studies of the glucocorticoid-binding protein from thymocytes. I. Localization in the cell and some properties of the protein. Biochim. biophys. Acta (Amst.) **214**, 520—532 (1970).

SCHULTZ, J.: Genes, differentiation and animal development. Brookhaven Symp. Biol. **18**, 116—147 (1965).

SERFLING, E., PANITZ, R., WOBUS, U.: Die experimentelle Beeinflussung des Puffmusters von Riesenchromosomen. I. Puffinduktion durch Oxytetracyclin bei *Chironomus tentans*. Chromosoma (Berl.) **28**, 107—119 (1969).

SHYAMALA, G., GORSKI, J.: Estrogen receptors in the rat uterus. Studies on the interaction of cytosol and nuclear binding sites. J. biol. Chem. **244**, 1097—1103 (1969).

SCHMUCKLER, E. A., KOPLITZ, M.: The effects of carbon tetrachloride and ethiamine on RNA synthesis in vivo and in isolated rat liver nuclei. Arch. Biochem. Biophys. **132**, 62—79 (1969).

SWIFT, H.: Molecular morphology of the chromosome. In: The Chromosome. In vitro, Vol. 1, pp. 26—49. Baltimore: Williams and Wilkins Co. 1965.

TARTOF, K. D., PERRY, R. P.: The 5S RNA genes of *Drosophila melanogaster*. J. molec. Biol. **51**, 171—183 (1970).

THOMSON, J. A., ROGERS, D. C., GUNSON, M. M., HORN, D. H. S.: Developmental changes in the pattern of cellular distribution of exogenous tritium-labelled crustecdysone in larval tissues of *Calliphora*. Cytobios. **6**, 79—88 (1970).

TOFT, D., GORSKI, J.: A receptor molecule for estrogens: Studies using a cell free system. Proc. nat. Acad. Sci. (Wash.) **55**, 1574—1581 (1966).

TSUZUKI, I.: Labeled RNA in isolated rat spleen nuclei. Exptl. Cell Res. **58**, 431—434 (1970).

— NAORA, N.: Protein synthesis in isolated rat spleen nuclei. Biochim. biophys. Acta (Amst.) **169**, 550—552 (1968).

WASSERMAN, M.: Cytological and phylogenetic relationships in the repleta group of the genus *Drosophila*. Proc. nat. Acad. Sci. (Wash.) **46**, 842—859 (1960).

WEIRICH, G., KARLSON, P.: Distribution of tritiated ecdysone in salivary glands and other tissues of *Rhynchsciara* and *Chironomus* larvae. An autoradiographical study. Wilh. Roux' Arch. **164**, 170—181 (1969).

WIMBER, D. E., STEFFENSEN, D. M.: Localization of 5S RNA genes on *Drosophila* chromosomes by RNA-DNA hybridization. Science **170**, 639—641 (1970).

Balbiani Ring Activities in Acricotopus lucidus

Reinhard Panitz

Zentralinstitut für Genetik und Kulturpflanzenforschung der Deutschen Akademie der Wissenschaften, Gatersleben, DDR

I. Introduction

The normal course of somatic cell division is such that all cells of a multicellular organism should possess the same genetic constitution. This fact seems to be in contradiction to the functional diversity among cells of higher organisms. The nowadays generally accepted explanation of this apparent paradox lies in the assumption that cell differentiation arises by differential activation of the genome, i.e. by activation and/or inactivation of different sets of genes. Convincing evidence to support this view comes from the study of the puffing phenomenon in Dipteran giant chromosomes. Puffs have been shown to be specific in relation to both time and developmental stage. BEERMANN's (1952) interpretation that puffing in polytene chromosomes is the visible manifestation of local gene activity has become well established in the last 20 years from numerous cytological and biochemical findings in several species of Diptera. Moreover, giant chromosomes provide a suitable model object for studying the regulation of gene activity at the chromosomal level, especially in relation to the topography of the chromosome.

A favorable opportunity to analyse questions dealing with the regulation of puffing and its bearing on cell function is offered by the midge *Acricotopus lucidus* (Diptera, Chironomidae, subf. Orthocladiinae). The salivary gland of *Acricotopus* has a complex structure showing extreme differences in cell function which are reflected in a high level of intragland variation of the Balbiani ring pattern, not found in other Chironomids.

II. Balbiani Ring Patterns in the Different Cell Types of the Salivary Gland

The salivary gland of *Acricotopus* forms a tripartite structure consisting of the large main lobe, the side lobe, and the anterior lobe consisting of 11—13 cells only. The three giant chromosomes exhibit in each cell type a unique puffing pattern, the specificity of which is principally determined by the formation of two to three large Balbiani rings (MECHELKE, 1953, 1962). On the whole there are six loci which regularly form a Balbiani ring (Fig. 1). Characteristic for the main and side lobes are the Balbiani rings BR-1 and BR-2, which are absent in the anterior lobe. In this part of the gland the only Balbiani rings characteristic of this cell type, BR-3, BR-4 and a further nucleolus-associated Balbiani ring (BR-7), are activated. Additionally, Balbiani ring BR-6 is formed in the side lobe.

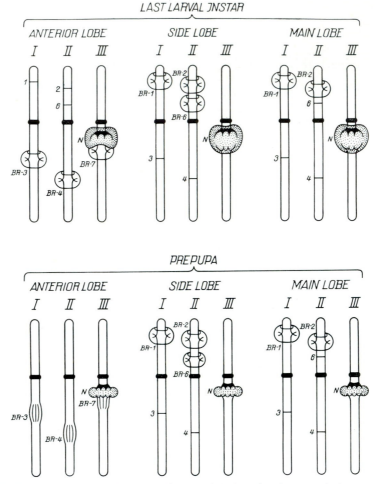

Fig. 1. Balbiani rings of the 3 different cell types of the larval and prepupal *Acricotopus* salivary gland. BR-1 to BR-7: the different Balbiani rings, N: nucleolus

A. The Balbiani Rings of the Main and Side Lobes

BR-2: BR-2, located in the short arm of chromosome II, is the largest Balbiani ring of the salivary gland. In extreme cases, its area may be seven times that of the chromosome diameter, which almost corresponds to the size of the nucleolus (Fig. 2). In comparison with the homologous, non-modified chromosomal segment from the anterior lobe, its origin was traced back to a region of only 1—2 bands (Mechelke, 1953).

BR-1: The second Balbiani ring, specific for the main and side lobes, is BR-1, located in the short arm of chromosome I. Its place of origin is most probably a

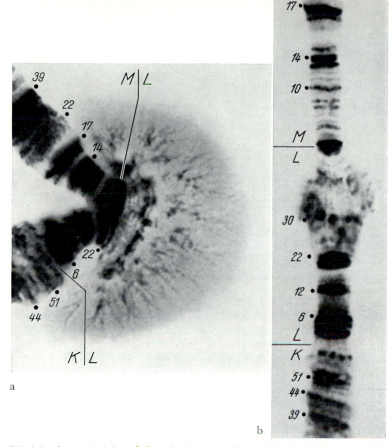

Fig. 2. BR-2 in the main lobe of the *Acricotopus* salivary gland. a fully expanded BR-2 in the larval salivary gland, b homologous chromosome segment in a young pupa with regressed BR-2. (After MECHELKE, 1962)

single thick band (MECHELKE, 1953). BR-1 is peculiar in that its size may vary in connection with a subterminal inversion in which it is included, i. e. it is dependent on its position in the chromosome (Fig. 3). In the homozygous type A (standard) and also in the heterozygous inversion (type B) BR-1 is always smaller than in the homozygous type C. As shown by a comparison of the area quotient BR-1/BR-2 of structural types A (0.56) and C (1.13), BR-1 reaches at least the size of BR-2 in type C (WOBUS et al., 1971). A similar case of position effect in the anterior lobe of the *Acricotopus* salivary gland was reported by MECHELKE (1960) and concerns the region A 42/45[1] included in the same inversion and separated by a segment of about 20 bands from BR-1. While this locus is always inactive in type A, it forms a small puff in special populations with type B, and a medium-sized. Balbiani ring (BR-5) in type C.

---

1 The marking of bands follows the chromosome map prepared by MECHELKE (unpublished).

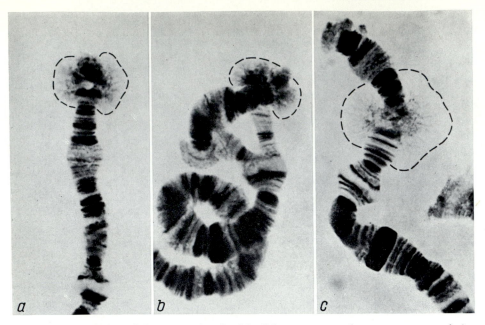

Fig. 3. Different size of BR-1 associated with different structural rearrangements of the short arm of chromosome I. a rearrangement A, b rearrangement B (heterozygous inversion), c rearrangement C. (After Wobus et al., 1971)

This situation is similar to that of BR-1 which also exhibits its maximal expansion in type C.

BR-6: A further Balbiani ring, specific for the side lobe only, is BR-6, located in the short arm of chromosome II proximal to BR-2, the size of which it rarely attains.

All Balbiani rings are completely unfolded during the larval phase of development. In the early prepupa, the volume of BR-6 decreases without complete regression. Only with the onset of histolysis of the salivary gland in the late prepupa all Balbiani rings become inactivated.

The single nucleolus of *Acricotopus* is formed by a heterochromatic region of chromosome III (Fig. 1). During the prepupal stage the nucleolus starts shrinking in all cell types of the salivary gland until it totally disappears in the old prepupa.

B. The Balbiani Rings of the Anterior Lobe

The anterior lobe occupies a special position in the salivary gland, possessing 3 Balbiani rings not found in the other lobes of the gland.

BR-3: This Balbiani ring is located in the middle of the long arm of chromosome I. According to Mechelke (1953), it originates from a segment of 3 bands of equal thickness. A unilocal origin is not ruled out. A distal transposition of BR-3 by a shift in combination with a complex structural rearrangement of the long arm of chromosome I does not affect its size (Wobus et al., 1971). BR-3 regularly disappears in the course of late prepupal development.

BR-4: The segment forming BR-4 is located in the long arm of chromosome II between bands P50 and P78. In the late larva, the area of BR-4 somewhat exceeds that of BR-3. BR-4 shows remarkable activity during the development from late larva to prepupa (MECHELKE, 1961). Its activity zone (region of maximal unfolding) is not stationary but shifts from distal to proximal over a range of about 15 bands during this time interval until it completely disappears in the late prepupa. This process was interpreted by MECHELKE as successive activation of different subunits (bands) of a complex gene locus. However, the activation of 2 distinct loci independently of each other is also possible. This idea finds support in the observation that in the case of ecdysone-induced inactivation of BR-4 (cf. p. 222) a self-dependent puff may occasionally be formed in the proximal subunit of BR-4 but separated from the distal subunit by several bands (PANITZ, 1964a).

BR-7: Comparing the nucleolus structure in the different lobes of the salivary gland, MECHELKE (1953) observed a structural peculiarity of the nucleolus from the anterior lobe, indicating the existence of an additional nucleolus-associated functional structure. The "nucleolus" in the anterior lobe consists of 2 distinct parts, the phase boundary of which is marked by numerous prick-like chromatin fragments (cf. Fig. 26 in MECHELKE, 1953). MECHELKE was unable to decide whether the nucleolus in the anterior lobe is associated with a Balbiani ring or with an accessory nucleolus. The absence of the typical ring-shape argued against the existence of a Balbiani ring, and the cell specificity of its activation against an accessory nucleolus. However, the discovery of a structural rearrangement resulting in a spatial separation of both functional structures solved this problem (PANITZ, unpublished). Both functional structures originate from a heterochromatic complex of bands which has been resolved into at least 7 bands (W11—W17) in analysis of animals possessing a functional heterozygosity of the nucleolus (PANITZ, 1960a). The region between W14 and W16 is considered the nucleolus organizer, while the additional functional structure was found to be formed by the double band W16/17 (Fig. 4). The two functional structures are separated from each other by one of the two inversions, the distal limit of which is located between W15 and W16/17. The resulting consequences for

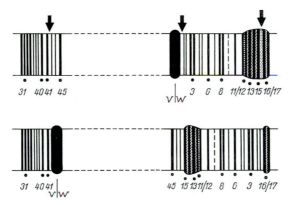

Fig. 4. Subdivision by inversion of the heterochromatic organizer region of the nucleolus and BR-7 (W 11/12-W 16/17) in chromosome III. Above: original arrangement. The limits of the two independent inversions are marked by arrows. Below: inversion-homozygous chromosome segment. V/W: centromere

animals with heterozygous inversions are illustrated in Fig. 5. While in the main and side lobes a normal-sized nucleolus and a smaller inverted one are formed, in the anterior lobe the double band W16/17 is activated in addition to a large Balbiani ring (BR-7). In all probability, only W17 is responsible for the formation of BR-7.

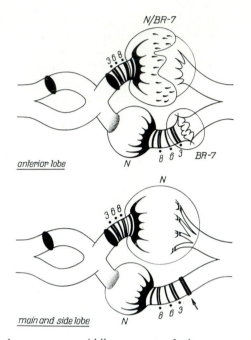

Fig. 5. The inversion-homozygous middle segment of chromosome III and the consequences for the activation of the nucleolus and BR-7 in different cell types of the salivary gland. Above: separate activation of the nucleolus and BR-7 in the anterior lobe. Below: the homologous chromosome segment in the main and side lobes without activated BR-7 region

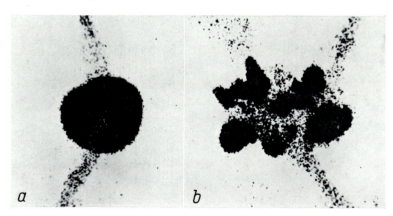

Fig. 6. Labeling differences of the nucleolus complex in the main lobe a) and the anterior lobe b) of the *Acricotopus* salivary gland. In the anterior lobe the weakly labeled BR-7 is located in the middle part of the complex and covered by the disrupted nucleolus. Autoradiograph after ³H-uridine incorporation

In the case of fully expanded BR-7, the band W16 is completely split up, forming numerous prick-like chromatin bits arranged at the phase boundary of both functional structures. There are two variants for the spatial arrangement of nucleolus and Balbiani ring: either the nucleolus extends into the Balbiani ring, deforming the latter into a hood-like structure covering the nucleolus (MECHELKE, 1953), or BR-7 keeps the central position, resulting in a fragmentation of the nucleolus body. The latter case is the most frequent one and is strikingly demonstrated by radioautography following ³H-uridine incorporation (Fig. 6).

III. Balbiani Rings and RNA Synthesis

Autoradiographic studies on the RNA synthesis of giant chromosomes in general have shown that the potential intensity of ³H-uridine incorporation corresponds to the degree of unfolding of a single chromomere (puff size). According to the autoradiographic experiments of PELLING (1964), the predominant sites of chromosomal RNA synthesis are the Balbiani rings (with the exception of the nucleolus) showing a rate of RNA synthesis 1 to 2 powers of ten higher than that of normal-sized puffs. Even different-sized Balbiani rings from one cell exhibit size-dependent differences of ³H-uridine incorporation independently of their actual rate of labeling. This finding was established by analysing pulse-labeled RNA extracted from single chromosomes or Balbiani rings of *Ch. tentans*, both *in vivo* (PELLING, 1970) and *in vitro* (DANEHOLT et al., 1969). According to a hypothesis of MECHELKE (1959), a direct correlation exists between the DNA content of a band and its potential unfolding. This idea has been confirmed by cytophotometrically ascertaining the DNA content of distinct Feulgen-stained bands of *Acricotopus* salivary gland chromosomes (MECHELKE, unpublished). Since the transcriptional activity of a band may be limited by its thickness or DNA content, small puffs might be formed by small, DNA-poor bands, whereas Balbiani rings should originate from thick, DNA-rich bands. In other words, the size of the puff seems to be to a certain degree a measure of the potential transcriptional activity of a gene locus.

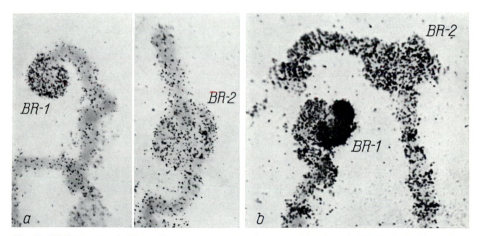

Fig. 7. Differences of ³H-uridine incorporation between BR-1 and BR-2 of the main lobe of the *Acricotopus* salivary gland. a weak labeling, b maximal labeling of both Balbiani rings

The validity of this view, however, is weakened by autoradiographic investigations of the Balbiani rings of *Acricotopus* (Panitz, Serfling, and Wobus 1972). There exist striking differences of labeling between individual Balbiani rings of the salivary gland which do not correlate with their area ratio. These differences are especially conspicuous between BR-1 and BR-2 in the main and side lobes of the salivary gland (Fig. 7).

In order to obtain comparable values for the degree and process of labeling at BR-1 and BR-2, silver grain counting was performed on ^{3}H-uridine autoradiographs. The maximum count yield over a Balbiani ring was found to be 30 grains per 10 μm^2. As the reference value, total labeling (100%) was used with a minimum grain density of 54.4 grains per 10 μm^2, calculated on the basis of an average grain diameter of 0.6 μm. Since this value is valid only for the ideal case of uniform grain distribution, the true number of grains necessary for total labeling may be somewhat higher.

A comparison of grain numbers obtained after incubation of salivary glands in haemolymph containing ^{3}H-uridine yielded a lower grain density for the larger BR-2 than for BR-1. Total labeling of BR-2 has never been found, not even when BR-2 in combination with structural type C of BR-1 exhibited a relatively smaller area than BR-1 (cf. p. 211). In a comparison of the labeling at BR-2 with different labeling classes of BR-1, BR-2 always showed lower grain densities independently of the actual labeling at BR-1. In this context, label differences of both Balbiani rings increased with increasing label density of BR-1 and reached the maximum with the total labeling of BR-1 (cf. Table 1).

Table 1. Comparison of ^{3}H-uridine incorporation (mean value $\pm$ standard error) within BR-1 and BR-2 for different classes of initial labeling of BR-1. n_1: number of animals, n_2: number of Balbiani rings examined

BR-1					BR-2			Difference in labeling between BR-1 and BR-2 (%)
n_1	Labeling classes		n_2	grains/10 μm^2 %	n_2	grains/10 μm^2 %		
3	weak labeling	(I)	15	6.7 $\pm$ 0.003	12.2	10	4.8 $\pm$ 0.9	7.7	4.5
7	mid labeling	(II)	36	13.8 $\pm$ 0.8	24.3	32	9.4 $\pm$ 1.1	17.3	8.0
5	strong labeling	(III)	22	20.0 $\pm$ 1.3	36.3	19	12.0 $\pm$ 2.4	22.0	14.3
4	total labeling	(IV)	20	54.4[a]	100	12	17.9 $\pm$ 2.2	32.8	67.2

[a] Calculated on the basis of an average grain diameter of 0.6 μm and uniform grain distribution.

This result raises the question as to what causes the consistently lower incorporation in the case of the larger BR-2. In general, two alternative interpretations are possible: 1) the potential capacity of RNA synthesis of BR-2 is related to its size, i.e. BR-2 reaches at least the activity of BR-1, or 2) no conformity exists between Balbiani ring size and rate of RNA synthesis, which would then really correspond to the low labeling level of BR-2. In the first case, there exists a contradiction between the high capacity for RNA synthesis and the low intensity of incorporation, in the second case, a contradiction between the extreme expansion of the Balbiani ring and its slight capacity for RNA synthesis. In the first case one might explain the different labeling behaviour by the assumption that, in contrast to BR-1, BR-2 is producing an eytremely uridine-poor RNA. From experiments of EDSTRÖM and BEERMANN (1962), the RNA of *Chironomus tentans* Balbiani rings is known to differ in base composition. These differences are also demonstrable microautoradiographically by the use of different ³H-labeled nucleosides (BEERMANN, personal communication). In order to clarify this question, a comparison was made of the incorporation of ³H-uridine, ³H-cytidine, ³H-adenosine and ³H-guanosine into BR-1 and BR-2 of *Acricotopus*, respectively. The experiments confirmed the already known labeling situation, i.e. essentially weaker labeling of BR-2.

After ³H-Cytidine incorporation in some cases, however, equally weak and total labeling of BR-2 was observed (the latter in connection with total labeling of BR-1), though it was never obtained with ³H-uridine (Fig. 8). This preliminary result may hint at the occurrence in BR-2 of an independent synthesis in time of two RNA fractions of different base composition. However, there is no experimentally founded basis for this view, which is based on observations in other objects, although the possibility that the DNA of one Balbiani ring may contain information for more than one polypeptide (GROSSBACH, 1969; WOBUS, PANITZ, and SERFLING, 1970) cannot be excluded.

In this context an interesting relationship exists with respect to a correlation in the activity of BR-2 in the main and side lobes of the *Acricotopus* salivary gland and the occurrence of proline hydroxylation (synthesis of an OH-proline-containing secretion protein) hinting at a causal relationship between the two events (BAUDISCH and PANITZ, 1968). The synthesis of either an OH-proline- or proline-rich protein would require a cytidine-rich RNA on the basis of the code for proline. It is tempting

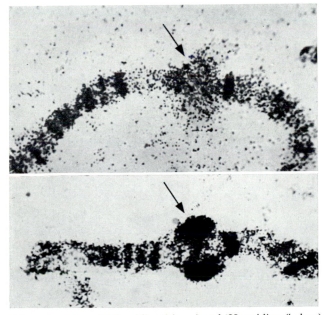

Fig. 8. Maximal incorporation of ³H-uridine (above) and ³H-cytidine (below) into the BR-2 of the main lobe of the salivary gland

to interpret this result as meaning that a cytidine-rich RNA is discontinuously released from BR-2, controlling the synthesis of an OH-proline and/or proline-rich protein.

The second interpretation of the weak labeling of BR-2 is based on the assumption that the lack of conformity between Balbiani ring size and RNA synthesis is due to a low synthetic capacity. We tried to test this possibility by autoradiographic examination of the kinetics of RNA synthesis and RNA turnover. As a measure of the rate of RNA synthesis, we used the time interval necessary to reach the Balbiani ring specific maximum of labeling. A comparison of the percentages of maximally labeled BR-1 and BR-2 with increasing incubation time, showed that BR-2 on average needs a longer time interval to attain its labeling maximum than BR-1. While after a 1-h incubation 35% of BR-1 exhibited maximal labeling, and after 4 hrs 82%, the corresponding values obtained for BR-2 were found to be 12 and 35%, respectively. From these results a lower rate of RNA synthesis of BR-2 might be inferred. However, for a definitive statement of this type, a knowledge of RNA turnover (retention period of RNA within the Balbiani ring) is necessary, because it cannot be excluded that in the case of BR-1 a high rate of RNA synthesis is simulated by long storage of the newly synthesized RNA within the Balbiani ring. Berendes (1968a) provided data showing that RNA synthesized in puffs of *Drosophila hydei* does not appear to be released immediately after its synthesis but is stored for about 20 min. On the basis of a different retention period of the newly synthesized RNA, the different labeling

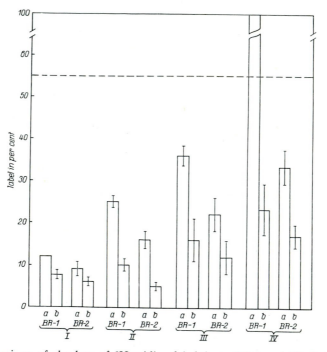

Fig. 9. Comparison of the loss of ³H-uridine label from BR-1 and BR-2 after different initial labeling (labeling classes I—IV, cf. table). a control gland (initial labeling), b experimental gland. The degree of labeling of the Balbiani rings is expressed as a percentage of total labeling (54.4 grains/10 μm²). The broken line indicates the limit of the maximum count yield

behaviour of BR-1 and BR-2 might be understood. In order to test this assumption, RNA release of both Balbiani rings was checked by measuring the loss in vitro of label after inhibition of RNA synthesis by actinomycin D. For this purpose, both salivary glands of larvae were incubated for 3—4 hrs in haemolymph containing ³H-uridine, after which one gland was fixed as a control and the other one postincubated in actinomycin-containing haemolymph without ³H-uridine. The results obtained are shown in Fig. 9. In this diagram, the release of BR-2 RNA was arranged according to different label classes of BR-1 (cf. Table). A comparison of the different classes exhibits in most cases a lower turnover rate for BR-2 than for BR-1. It is also noticeable that the turnover rate of both Balbiani rings increased with increasing initial labeling. The explanation for this is that a decreasing amount of unlabeled RNA is leaving the Balbiani rings until labeling saturation is reached. Consequently, the turnover maximum is ascertained only when the Balbiani ring specific maximum of labeling is reached. This turned out to be about 75% of the labeled RNA per hr for BR-1 and about 50% for BR-2. The intensity of RNA turnover became most obvious after doubling the time of postincubation up to a maximum of 2 hrs. The increase of RNA release (loss of label) was found to be about 10% (total 85%) of the initial labeling for BR-1 and about 20% (total 60%) for BR-2. This indicates that, at least in case of BR-1, most of the newly synthesized RNA leaves the chromosome within 1 h. Similar results were obtained with *Chironomus tentans*. *In vitro*, already after 30 min the heterogeneous RNA of chromosome IV (which possesses all 3 Balbiani rings) shows a complete turnover, while the turnover time is somewhat longer for the chromosomes not carrying Balbiani rings (DANEHOLT et al., 1969).

These data on the kinetics of labeling clearly demonstrate different capacities of RNA synthesis for BR-1 and BR-2 inversely proportional to their size and meaning that the larger BR-2 possesses a lower synthetic capacity than the smaller BR-1. This raises the question as to the cause of the contradiction between maximal unfolding and low capacity for RNA synthesis. It is conceivable that only a small part of the uncoiled (and redundant?) Balbiani ring DNA is transcribed, while the major part fulfills other functions (cf. BEERMANN, 1965). This would mean that, in the case of BR-2 of *Acricotopus* and in contrast to BR-1, less DNA is available for transcription. This assumption is supported by the observation of BERENDES (1968a) that in *Drosophila hydei* the de-novo synthesis of RNA in an induced puff is restricted to a small segment of the extended DNA. Another indication for this view is the discovery of the existence of the multiple spacer segment within the redundant rDNA of *Xenopus laevis* which is transcribed to one tenth of its length only (MILLER and BEATTY, 1969; DAWID, BROWN, and REEDER, 1970). Still other explanations could be sought in the assumption of an extremely low molecular weight RNA of BR-2 or of a different Balbiani ring-specific concentration of RNA polymerase. That such differences in concentration may be the reason for different rates of RNA synthesis, was demonstrated by HAMILTON (1968) and LINDELL (cited in TOMKINS and MARTIN, 1970).

IV. The Regulation of Balbiani Ring Activity

Puffing pattern changes occur regularly during the postembryonic development of Diptera (BEERMANN, 1952; MECHELKE, 1953; BREUER and PAVAN, 1955; BECKER, 1959; BERENDES, 1965; ASHBURNER, 1967). Especially during the moulting periods and prior to pupation, drastic changes in the pattern of tissue-specific puffing activity occur in temporally regulated sequences. The simplest explanation for this phenomenon is that puffing pattern changes reflect the qualitative and quantitative regulation of the transcriptional activity of the genome at the chromosomal level. If so, similar changes should obviously be assumed to occur at the level of translation.

However, nearly all experiments carried out to prove such changes at the protein level have been unsuccessful. A comparison of soluble proteins of the salivary gland cells of *Acricotopus* and two other Chironomids at different developmental stages did not yield any striking differences in the protein patterns (Wobus, 1970). Furthermore, the attempt to influence protein synthesis of intermoult larvae of *Chironomus* by long-time inhibition of RNA synthesis was without success (Clever, Bultman, and Darrow, 1969; Doyle and Laufer, 1969; Wobus et al., 1972). A few hours after moulting to the 4th larval instar, protein synthesis was partially blocked by actinomycin D, but no changes in puffing could be observed at this time (Clever, Storbeck, and Romball, 1969). Therefore, we may conclude that puffing pattern changes do not immediately cause concurrent changes in cellular function at the level of proteins. The latter changes must be explained on the basis of stable messengers which are translated at different times after synthesis. On the other hand, an additional Balbiani ring in a *Ch. thummi* strain could in no way be correlated with the synthesis of an additional protein (Wobus, Serfling, and Panitz, 1971). Probably, puff RNA is only partly used in translation, a view supported by the discovery that the greater part of the nuclear RNA turns over very rapidly without leaving the nucleus (Daneholt and Svedhem, 1971).

Autoradiographic experiments proved the hypothesis that changes in puffing pattern reflect changes in the pattern of RNA synthesis along the giant chromosomes. Although the existence of a puff is not an absolute indication of RNA synthesis, as shown in experiments on ^{3}H-uridine incorporation within the Balbiani rings of *Ch. tentans* (Pelling, 1964) and *Acricotopus* (Panitz, unpublished), the belief seems to be justified that changes in the puffing pattern can be interpreted as the reflection of the selective regulation of genomic activity and thus may be used as a model in studies on gene regulation.

A. Development-specific Changes of BR-3 and BR-4

As mentioned above, a striking feature of the *Acricotopus* salivary gland chromosomes consists in the development-specific formation of Balbiani rings. The most remarkable event in the development of the salivary gland is the inactivation of the Balbiani rings in the anterior lobe, while those of the other lobes remain active. To get an exact timetable of the activity behaviour of BR-3 and BR-4 during normal development, the salivary glands of single-bred larvae were dissected at different time intervals after the last larval moult, after which the degree of Balbiani ring activity was determined. In dependence on the breeding temperature, the Balbiani ring regression starts at different points of time after moulting, proceeding continuously until the end of the prepupal stage. Compared with BR-3, BR-4 shows during this time a distinct delay which is maintained until the end of the prepupal phase (Fig. 10). As will be shown later on, this delay is not due to a shift of its zone of maximum unfolding, but rather to a locus-specific feature.

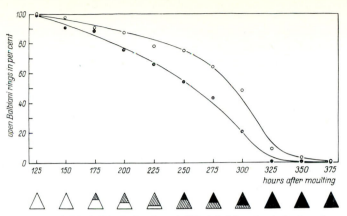

Fig. 10. Regression of BR-3 (black dots) and BR-4 (white dots) during the prepupal stage at 10° C. The activity of the Balbiani rings is compared with the degree of carotenoid accumulation in the anterior lobe of the salivary gland symbolized by triangles. (After PANITZ, 1964b)

Parallel with the regression of BR-3 and BR-4, a yellow or brown colour of the secretion of the anterior lobe occurs with slight retardation (MECHELKE, 1953; PANITZ, 1964a). According to the investigations of BAUDISCH (1963), the yellow pigment results from the accumulation of carotenoids, mainly β-carotene. The correlation between these events is very striking: both the regression of the Balbiani rings and the accumulation of the yellow pigment begin at the tip of the anterior lobe and proceed up to the base of the lobe in the course of prepupal development (Fig. 10). The parallel of both events is always present; thus, regressed Balbiani rings were never found in any animal without the yellow secretion and vice versa. It is not clear whether there is any causal relationship between the two processes. It is certain that the anterior lobe was found to be completely inactive in secretion, according to an electron microscopic analysis of the cytoplasmic structures (DÖBEL, 1968, and unpublished).

B. Hormone-induced Changes of BR-3 and BR-4

The strict developmental specifity of BR-3 and BR-4 activity, especially the temporal conformity between Balbiani ring regression and prepupal development, suggested a control of Balbiani ring activity by development-controlling factors. Experiments started some time ago to clarify this question led to the following main results (PANITZ, 1960b, 1964a): 1. the inactivation of BR-3 and BR-4 is inducible (without influencing the activity of the other Balbiani rings) a) by transplantation of a larval salivary gland into a prepupa of *Acricotopus*, *Chironomus* and *Calliphora*, b) by incubation of a larval gland in larval haemolymph with added ring glands of *Calliphora*. 2. Inactivation is impossible a) by transplantation of a larval salivary gland into an *Acricotopus* larva, and b) by incubation of a larval gland in larval haemolymph.

The fact that the prepupa-specific puffing pattern is inducible by an alteration in the internal milieu indicates the significance of hormonal factors for Balbiani ring activity. This assumption was supported by the discovery that a short-term inactivation of BR-3 and BR-4 takes place also during the last larval moult, but only with respect to a restricted number of Balbiani rings (Panitz, 1964a). This assumption has in the meantime been confirmed. The inactivation of BR-3 and BR-4 can be induced experimentally by injection of the hormone ecdysone into larvae at an intermoult stage (Panitz, Wobus and Serfling, 1972), or by incubating salivary glands in a medium containing ecdysone (Panitz, unpublished). In the range of physiological concentrations (cf. p. 223) the injection of the hormone leads to a regression of BR-3 and BR-4, as is typical of normal development. The first significant reaction of BR-3 occurred after about 15 min, of BR-4 somewhat later. With regard to its structure, BR-3 showed no difference from normal development. However, in the case of BR-4, the typical shifting of its zone of maximum unfolding was not observed (cf. p. 213). The same cytological results were obtained by injecting ecdysterone and inokosterone. This is in agreement with Berendes (1968b) and Ashburner (1970), who were able to induce ecdysone-dependent puffs in *Drosophila hydei* and *D. melanogaster*, respectively, by means of ecdysterone. Compared with ecdysone, ecdysone analogues proved less than half as effective when applied in equal doses (2×10^{-1} µg). The weakest effect was exerted by inokosterone.

The accumulation of the secretion in the anterior lobe containing the yellow pigment was not observed after ecdysone injection. This, however, does not exclude a causal relationship between the hormone and the Balbiani ring activity, or the alteration of the secretion. The rapid course of the induced processes would not leave sufficient time to allow the accumulation of carotenoids, which most probably are obtained from an exogeneous source. After hormone injection a time span of 1—2 hrs is necessary for the complete inactivation of the Balbiani rings, while the same process during normal development needs several days. Thus, the inactivation of BR-3 and BR-4 during slow development (at 10° C) needs 10 days on an average, and the accumulation of the carotenoids 6 days, while the corresponding data for fast development (at 24° C) were found to be 4 and 2 days, respectively. Berendes (1965) found in *Drosophila hydei* a correlation between the appearance of a puff and a structural reorganization of the cytoplasm resulting in a specific secretory component (mucopolysaccharide). This reaction chain is experimentally inducible by the injection of ecdysone and suggests a hormonal control (Poels, 1970). Although electron microscopic investigations on the cytoplasmic structures of ecdysone-treated salivary glands of *Acricotopus* are lacking, it can be supposed on the basis of results obtained from normal development (Döbel, 1968) that in the present case the secretory activity of the anterior lobe is controlled by ecdysone, most probably via the inactivation of BR-3 and/or BR-4.

Previous results have already shown that the intensity of Balbiani ring inactivation depends on the concentration of the applied factor (Panitz, 1964a). A comparison of different ecdysone concentrations, confirmed this statement in principle, but with the restriction that the relationship between hormone doses and effect is not linear over the whole range of concentrations studied (Fig. 11). Above the concentration of 2×10^{-5} µg an optimum range of reaction exists, in which a subsequent increase of the ecdysone concentration leads to a higher effect than below this limit. While Clever (1963) reported a linear dose-response relationship in *Chironomus tentans*, Berendes (1967) found in *Drosophila hydei* a situation as complex

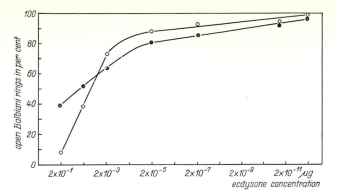

Fig. 11. Relationship between different concentrations of ecdysone and the degree of regression of BR-3 (black dots) and BR-4 (white dots) 1 hr after injection. (After PANITZ et al., 1972)

as in *Acricotopus*, indicating that gene reactions are not always simple dose-effect relationships.

Another significant feature of the dose-effect curves consists in the different sensitivity of the two Balbiani rings to equal hormone concentrations. In the range of higher concentrations (2×10^{-1} to 0.5×10^{-2} µg), BR-4 exhibits a stronger reaction than BR-3, while the situation in the lower concentration range (right to the intersection point) is exactly reversed. The reaction of BR-3 and BR-4 illustrated on the right side of the diagram is almost identical with that already described for normal development (cf. Fig. 10). On the other hand, the reaction on the left of the intersection is typical only for experimental conditions using non-physiological hormone concentrations. From these results, based on an average larval fresh-weight of 2.0 mg, one may infer a maximal ecdysone titer of 0.5×10^{-3} µg in the old *Acricotopus* prepupa.

An interesting question is, whether BR-3 and BR-4 react directly on ecdysone or via another control system. CLEVER (1964) has shown that the induction of two puffs in *Chironomus tentans* is followed 10—48 hrs later by a sequential activation of other puffs. The appearance of these puffs can be stopped by inhibition of protein synthesis; this led CLEVER to the conclusion that the early puffs control the activation of the later ones via protein synthesis. In *Acricotopus* it has so far been impossible to detect any reproducible puff reaction immediately after hormone injection yet before the onset of the Balbiani ring regression. On account of the short time interval of 15 min between hormone injection and the first visible reaction of BR-3, the existence of such a control system is most unlikely. Inhibition of protein synthesis by treatment of *Acricotopus* larvae with cycloheximide prior to ecdysone injection revealed no difference between cycloheximide-treated and control larvae, thus excluding any such hypothetical control system (PANITZ, WOBUS and SERFLING, 1972).

C. Experimental Modification of Balbiani Ring Activity

A large variety of agents is known to influence significantly the puffing pattern in polytene chromosomes. These inducible changes fall into 3 categories without clearcut boundaries between them:

> changes in puff size,
> induction of puffs unknown or known in normal development,
> inactivation of puffs.

The mechanism by which different agents cause changes in the puffing pattern is assumed to be different for each of the 3 categories. Stimuli of the first category interfere with a regulatory mechanism already operating in the cell, while the de-novo induction of puffs may be considered as gene derepression. In most cases, a single agent induces a variety of puffing pattern changes, suggesting that it influences an unspecific regulatory component common to all reacting loci. The reaction of only one puff or Balbiani ring, however, must be understood as indicating interference with a locus-specific regulatory mechanism. Such cases would be important not only for studies on the problem of gene regulation, but they may also offer the opportunity to discover causal relationships between a puff or Balbiani ring and a distinct cell function. However, none of the many agents known to change puffing patterns has been reported to influence with certainty one locus only.

In treatments of *Acricotopus* larvae, gibberellins (A_3, A_4) cause the selective regression of only two Balbiani rings (PANITZ, 1967). If larvae are placed for 36—48 hrs in an aqueous solution of gibberellin A_3 (GA_3) of a concentration in the range 0.5 to 1.5 mg per ml, the BR-2 and BR-6 are nearly completely regressed, while the Balbiani rings in the anterior lobe of the gland (BR-3, BR-4, BR-7) remain unaffected. Only after prolonged treatment does a slight regression of BR-1 occur (about 10%). The total regression of BR-2 as well as the partial regression of BR-1 are completely reversible if the animals are replaced in fresh water. Autoradiographic experiments with ^{3}H-uridine demonstrated that treatment with GA_3 blocks RNA synthesis in BR-2, even if its expanded structure is preserved. This fact clearly indicates that structural regression is not an absolute prerequisite for the transcriptional inactivity of a Balbiani ring.

The mechanism of inactivation of BR-2 is completely unknown. It seems, however, that the mode of action of GA_3 is quite different from that of ecdysone, because no ecdysone-like effects were observed, as shown by the insensitivity of the ecdysone-controlled Balbiani rings of the anterior lobe. However, it may be possible for other insect species to transform gibberellin into the isoprenoid unit, which can then be used as ecdysone precursor. Thus, CARLISLE and coworkers (1963) found that GA_3 can induce additional moults in locusts after injection.

The long time interval necessary to induce the first visible signs of regression of BR-2 renders it unlikely that the plant hormone acts directly on the Balbiani ring. Therefore, the effect of GA_3 was tested in injection experiments (PANITZ, unpublished). First of all, there was a surprising result insofar as control injections with the solvent produced a reversible regression of BR-2 with a maximum of inactivation (about 80%) after 2 hrs (Fig. 12). The recovered Balbiani rings regain their original size only after some hours. Nothing is known about the causes of this short-term effect. The fact that BR-2 responds equally well after injection of haemolymph of larvae of the same physiological age excludes the assumption that the concentration of a potential control factor in the haemolymph is lowered. Wounding is perhaps the primary reason, since this treatment also alone results in a temporary inactivation

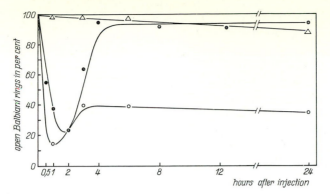

Fig. 12. The reaction of BR-1 (triangles) and BR-2 (white dots) on the injection of gibberellin A₃ (0.3 µg per animal). The effect of a gibberellin-free control solution is symbolized by black dots

of BR-2. In contrast with the control, injection of GA₃ exerts a strong and long-lasting inhibitory effect on the activity of BR-2, resulting in a nearly complete regression of the Balbiani ring structure. Due to interference with the control effect, a transitory regression maximum of about 85% occurs 1 h after GA₃ injection. The degree of regression depends on the batch used and on the concentration of GA₃. The first ascertainable effect is obtained with 0.3 µg per larva of the most effective batch of GA₃. This value approximately corresponds to the amount of 1.5 µg per larva which was found to induce puffs in *Drosophila hydei* (ALONSO, 1970).

The results reported may be interpreted as favouring a higher sensitivity of the BR-2 specific regulatory mechanism. Further experiments with other unspecifically acting agents seem to confirm this assumption. Treatment of larvae with actinomycin D results in the regression of all Balbiani rings, the degree of regression being dependent on the concentration, the temperature of treatment and on the Balbiani ring itself (SERFLING, 1968). In all cases, BR-2 is the first to react (Fig. 13). Similar effects were obtained after inhibition of protein synthesis by oxytetracycline or after treating the larvae with caffeine. In general, the sensitivity of the Balbiani rings of

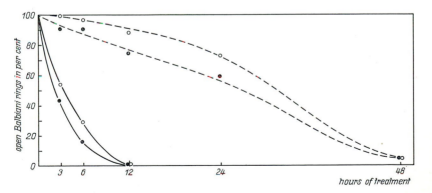

Fig. 13. Induced regression of BR-1 (white dots) and BR-2 (black dots) after treatment of larvae with 2 µg per ml (broken line) and 20 µg per ml (unbroken line) actinomycin D (After SERFLING, 1968)

Acricotopus can be said to decrease in the following order: BR-2 > BR-1 > BR-3 > BR-4. This sequence is not the same as the sequence of Balbiani ring regression during normal development (BR-3, BR-4, BR-2, BR-1).

References

Alonso, C.: The effect of gibberellic acid upon developmental processes in *Drosophila hydei*. Ent. exp. and appl. **14**, 73—82 (1971).

Ashburner, M.: Pattern of puffing activity in the salivary gland chromosomes of *Drosophila*. I. Autosomal puffing patterns in a laboratory stock of *D. melanogaster*. Chromosoma (Berl.) **21**, 398—428 (1967).

— Function and structure of polytene chromosomes during insect development. Advanc. Insect Physiol. **7**, 1—95 (1970).

Baudisch, W.: Chemisch-physiologische Untersuchungen an den Speicheldrüsenchromosomen von *Acricotopus lucidus*. 100 Jahre Landwirtschaftl. Institute d. Univ. Halle, pp. 152—159 (1963).

— Panitz, R.: Kontrolle eines biochemischen Merkmals in den Speicheldrüsen von *Acricotopus lucidus* durch einen Balbiani-Ring. Exp. Cell Res. **49**, 470—476 (1968).

Becker, H.-J.: Die Puffs der Speicheldrüsenchromosomen von *Drosophila melanogaster*. I. Beobachtungen zum Verhalten des Puffmusters im Normalstamm und in zwei Mutanten, *giant* und *lethal-giant-larvae*. Chromosoma (Berl.) **10**, 654—678 (1959).

Beermann, W.: Chromomerenkonstanz und spezifische Modifikation der Chromosomenstruktur in der Entwicklung und Organdifferenzierung von *Chironomus tentans*. Chromosoma (Berl.) **5**, 139—198 (1952).

— Operative Gliederung der Chromosomen. Naturwissenschaften **52**, 365—384 (1965).

Berendes, H. D.: Salivary gland function and chromosomal puffing pattern in *Drosophila hydei*. Chromosoma (Berl.) **17**, 35—77 (1965).

— The hormone ecdysone as effector of specific changes in the pattern of gene activities of *Drosophila hydei*. Chromosoma (Berl.) **22**, 274—293 (1967).

— Factors involved in the expression of gene activity in polytene chromosomes. Chromosoma (Berl.) **24**, 418—437 (1968a).

— The effect of ecdysone analogues on the puffing pattern of *Drosophila hydei*. Drosophila Inf. Serv. **43**, 145 (1968b).

— Induction and control of puffing. Ann. Embryol. Morphogen. Suppl. **1**, 153—164 (1969).

Breuer, M. E., Pavan, C.: Behaviour of polytene chromosomes of *Rhynchosciara angelae* at different stages of development. Chromosoma (Berl.) **7**, 371—386 (1955).

Carlisle, D. B., Osborne, D. J., Ellis, P. E., Moorhouse, J. E.: Reciprocal effects of insect and plant-growing growth substances. Nature (Lond.) **200**, 1230 (1963).

Clever, U.: Von der Ecdysonkonzentration abhängige Genaktivitätsmuster in den Speicheldrüsenchromosomen von *Chironomus tentans*. Develop. Biol. **6**, 73—98 (1963).

— Actinomycin and puromycin: Effects on sequential gene activation by ecdysone. Science **146**, 794—795 (1964).

— Bultman, H., Darrow, J. M.: The immediacy of genomic control in polytene cells. In: Hanly, E. W. (Ed.): Problems in Biology: RNA in Development, pp. 403—423. Salt Lake City, pp. 403—423, 1969.

— Storbeck, I., Romball, C. G.: Chromosome activity and cell function in polytene cells. I. Protein synthesis at various stages of larval development. Exp. Cell Res. **55**, 306—316 (1969).

Daneholt, B., Edström, J.-E., Egyházi, E., Lambert, B., Ringborg, U.: Chromosomal RNA synthesis in polytene chromosomes of *Chironomus tentans*. Chromosoma (Berl.) **28**, 399—417 (1969).

— Svedhem, L.: Differential representation of H RNA in nuclear sap. Exp. Cell Res. **67**, 263—272 (1971).

Dawid, I. B., Brown, D. D., Reeder, R. H.: Composition and structure of chromosomal and amplified ribosomal RNA's of *Xenopus laevis*. J. mol. Biol. **51**, 341—360 (1970).

Döbel, P.: Über die plasmatischen Zellstrukturen der Speicheldrüse von *Acricotopus lucidus* Kulturpflanze **16**, 203—214 (1968).

DOYLE, D., LAUFER, H.: Requirements of ribonucleic acid synthesis for the formation of salivary gland specific proteins in larval *Chironomus tentans*. Exp. Cell. Res. **57**, 205—210 (1969).

EDSTRÖM, J.-E., BEERMANN, W.: The base composition of nucleic acids in chromosomes, puffs, nucleoli and cytoplasm of *Chironomus tentans* salivary gland cells. J. Cell Biol. **14**, 371—379 (1962).

GROSSBACH, U.: Chromosomen-Aktivität und biochemische Zelldifferenzierung in den Speicheldrüsen von *Camptochironomus*. Chromosoma (Berl.) **28**, 136—187 (1969).

HAMILTON, T. H.: Control by estrogen of genetic transcription and translation. Science **161**, 649—661 (1968).

MECHELKE, F.: Reversible Strukturmodifikationen der Speicheldrüsenchromosomen von *Acricotopus lucidus*. Chromosoma **5**, 511—661 (1953).

— Beziehung zwischen der Menge der DNS und dem Ausmaß der potentiellen Oberflächenentfaltung von Riesenchromosomen-Loci. Naturwissenschaften **46**, 609 (1959).

— Strukturmodifikationen in Speicheldrüsenchromosomen von *Acricotopus* als Manifestation eines Positionseffekts. Naturwissenschaften **47**, 334—335 (1960).

— Das Wandern des Aktivitätsmaximums im BR$_4$-Locus von *Acricotopus lucidus* als Modell für die Wirkungsweise eine komplexen Locus. Naturwissenschaften **48**, 29 (1961).

— Spezielle Funktionszustände des genetischen Materials. Wiss. Konferenz Ges. dtsch. Naturf. u. Ärzte, Rottach-Egern, S. 15—29 (1962).

MILLER, O. L., BEATTY, B. R.: Visualization of nucleolus genes. Science **164**, 955—957 (1969).

PANITZ, R.: Gewebespezifische Manifestierung einer Heterozygotie des Nucleolus in Speicheldrüsenchromosomen von *Acricotopus lucidus*. Naturwissenschaften **47**, 359 (1960a).

— Innersekretorische Wirkung auf Strukturmodifikationen der Speicheldrüsenchromosomen von *Acricotopus lucidus*. Naturwissenschaften **47**, 383 (1960b).

— Hormonkontrollierte Genaktivitäten in den Riesenchromosomen von *Acricotopus lucidus*. Biol. Zbl. **83**, 197—230 (1964a).

— Experimentell induzierte Inaktivierung Balbiani-Ring bildender Gen-Loci in Riesenchromosomen. In: Struktur und Funktion des genetischen Materials. Erwin-Baur-Gedächtnisvorlesungen, Bd. **3**, S. 213—233 (1964b).

— Funktionelle Veränderungen an Riesenchromosomen nach Behandlung mit Gibberellinen. Biol. Zbl. **86** (Suppl.), 147—156 (1967).

— WOBUS, U., SERFLING, E.: The effect of ecdysone and ecdysone analogues on two Balbiani rings of *Acricotopus lucidus*. Exp. Cell Res. **70**, 154—160 (1972).

— SERFLING, E., WOBUS, U.: Autoradiographische Untersuchungen zur RNA-Syntheseleistung von Balbiani-Ringen. Biol. Zbl. **91**, 359—380 (1972).

PELLING, C.: Ribonukleinsäure-Synthese der Riesenchromosomen. Autoradiographische Untersuchungen an *Chironomus tentans*. Chromosoma (Berl.) **15**, 71—123 (1964).

— Puff RNA in polytene chromosomes. Cold Spring Harb. Symp. quant. Biol. **35**, 521—531 (1970).

POELS, C. L. M.: Time sequence in expression of various developmental characteres induced by ecdysone in *Drosophila hydei*. Develop. Biol. **23**, 210—225 (1970).

TOMKINS, G. M., MARTIN, D. W.: Hormones and gene expression. Ann. Rev. Genetics **4**, 91—106 (1970).

SERFLING, E.: Die Induktion funktioneller Veränderungen an Riesenchromosomen. Diploma Thesis, Martin-Luther-Universität Halle-Wittenberg (1968).

WOBUS, U.: Chromosomale Differenzierung und Proteinmuster bei Chironomiden. Ergebn. exp. Med. (Berl.) **2**, 23—26 (1970).

— PANITZ, R., SERFLING, E.: Tissue specific gene activities and proteins in the *Chironomus* salivary gland. Molec. gen. Genetics **107**, 215—223 (1970).

— POPP, SONJA, SERFLING, E., PANITZ, R.: Protein synthesis in the *Chironomus thummi* Salivary gland. Molec. gen. Genetics **116**, 309—321 (1972).

— SERFLING, E., PANITZ, R.: The salivary gland proteins of a *Chironomus thummi* strain with an additional Balbiani ring. Exp. Cell Res. **65**, 240—245 (1971).

— — BAUDISCH, W., PANITZ, R.: Chromosomale Strukturumbauten bei *Acricotopus lucidus* korreliert mit Änderungen im Proteinmuster. Biol. Zbl. **90**, 433—441 (1971).